Fundamentals of Ethics

FOR SCIENTISTS AND ENGINEERS

Fundamentals of Ethics

FOR SCIENTISTS AND ENGINEERS

Edmund G. Seebauer
University of Illinois at Urbana-Champaign

Robert L. Barry

New York • Oxford
OXFORD UNIVERSITY PRESS
2001

Oxford University Press

Oxford   New York
Athens   Auckland   Bangkok   Bogotá   Buenos Aires   Calcutta
Cape Town   Chennai   Dar es Salaam   Delhi   Florence   Hong Kong   Istanbul
Karachi   Kuala Lumpur   Madrid   Melbourne   Mexico City   Mumbai
Nairobi   Paris   São Paulo   Singapore   Taipei   Tokyo   Toronto   Warsaw

and associated companies in
Berlin   Ibadan

Copyright © 2001 by Oxford University Press, Inc.

Published by Oxford University Press, Inc.,
198 Madison Avenue, New York, New York 10016
http://www.oup-usa.org

Oxford is a registered trademark of Oxford University Press.

Library of Congress Cataloging-in-Publication Data

Seebauer, Edmund Gerard.
Fundamentals of ethics for scientists and engineers / Edmund G. Seebauer,
Robert L. Barry.
p.   cm.
Includes bibliographical references and index.
ISBN 0-19-513488-5
1. Science—Moral and ethical aspects.   2. Technology—Moral and ethical
aspects.   I. Barry, Robert Laurence.   II. Title.
Q175.35   .S44   2000
174'95—dc21   00-027879

9 8 7 6 5 4 3 2 1

Printed in the United States of America
on acid-free paper

Dedicated to the memory of
Aristotle (384–322 B.C.)
and
Thomas Aquinas (1225–1274)
who forged the best science
and the best philosophy of their day
into a coherent unity.

TABLE OF CONTENTS

Preface xiii

UNIT ONE FOUNDATIONAL PRINCIPLES

CHAPTER 1 APPROACHING THE SUBJECT OF ETHICS 3
An Example 3
The Importance of Ethics in Science and Engineering 4
Managing Ethical Discussion 6
Philosophy, Religion, and Ethics 6
The Existence of Right and Wrong 8
*Principle: Certain aspects of right and wrong exist
objectively, independent of culture or personal
opinion 8*
The Subject of Moral Analysis 9
The Role of Codes of Ethics 9
*A Real-life Case: Destruction of the Spaceship
Challenger 10*
Note 11
Problems 11

CHAPTER 2 THE PERSON AND THE VIRTUES 19
Developing a Model for the Person 19
Components of the Psyche 20
Limitations of the Model 21
Habits and Morals 22
The Four Main Virtues 22
*Principle: People should always decide and act
according to the virtues insofar as possible 24*
An Example 24
A Real-life Case: Toxic Waste at Love Canal 26
Notes 27
Problems 28

CHAPTER 3 ANALYZING EXTERIOR ACTS: SOME FIRST STEPS 25

Ethics as a Craft 35
Distinguishing Exterior and Interior Morality 36
Beginning Case Analysis 37
Event Trees 40
    A Real-life Case: Dow Corning Corp. and Breast
    Implants 42
Notes 43
Problems 43

CHAPTER 4 ANALYZING INTERIOR INTENTIONS: SOME FIRST STEPS 52

Describing Intention 52
The Importance of Intention 53
Effort and the Virtues 55
    Principle: People should try insofar as possible to
    continue to progress in the moral life 56
The Role of Benevolence 56
    A Real-life Case: The Tuskegee Syphilis Experiment 57
Note 58
Problem 58

SUMMARY 66

Some Words of Caution 67
Note 67

UNIT TWO RESOLVING ETHICAL CONFLICTS

CHAPTER 5 TOWARD A HIERARCHY OF MORAL VALUES 71

On Selecting Principles and Methods 71
Hierarchies of Values: Moral and Nonmoral 72
Line-drawing 73
An Example 75
Mathematical Analogies 76
Ranking the Virtues 78
    A Real-life Case: Scientific Tests Using Animals 78
Notes 80
Problems 80

CHAPTER 6 STARTING MORAL JUDGMENTS: EVALUATING
EXTERIOR ACTS 88

A Mathematical Analogy 88
An Example 90
    A Real-life Case: Chemical Disaster at Bhopal 96

Notes   97
Problems   98

CHAPTER 7   COMPLETING MORAL JUDGMENTS: THE DECISIVE
ROLE OF INTENTION   106
Evaluating Interior Goodness   106
An Example   107
Balancing Evaluations of Interior and
Exterior Goodness   110
The "Solomon Problem"   111
    Principle: The obligation to avoid what is bad
    outweighs the obligation to do what is good   111
Cooperating in the Evil of Others   111
    A Real-life Case: The Problem of Performance
    Evaluation—Grade Inflation   113
Notes   114
Problems   115

CHAPTER 8   MORAL RESPONSIBILITY   124
Factors Limiting Moral Responsibility   124
Degrees of Responsibility   125
An Example   126
The "Sainthood" and "Devil" Problems   128
    A Real-life Case: Responsibility in Software
    Engineering   129
Note   130
Problems   130

SUMMARY   138
Some Words of Caution   138

UNIT THREE   JUSTICE: APPLICATIONS

CHAPTER 9   TRUTH: PERSON-TO-PERSON   143
Truth in Actions   143
Truth in Words   144
Harm from Deception   144
Harm from Withholding Truth   145
Whistleblowing   146
Harm from Spreading Truth   147
Privacy   147
    A Real-life Case: Censorship of the Internet   148

Notes 150
Problems 150

CHAPTER 10 TRUTH: SOCIAL   157
Distinctions between Science and Engineering   157
Approach to Knowledge in Science   158
Recognition from Scientific Publication   159
Black and Gray in Scientific Practice   160
Approach to Knowledge in Technology   161
Intellectual Property   162
A Real-life Case: Copying Music Illegally Using the Internet   164
Notes 165
Problems 166

CHAPTER 11 FAIRNESS: PERSON-TO-PERSON   173
Conflict of Interest   173
Qualitative versus Quantitative Fairness   174
Credit or Blame in Team Projects   175
Authorship Questions   175
Fairness in Supervising   176
Fairness in Contracting with Clients   178
A Real-life Case: Problems with Peer Review   179
Notes 180
Problems 180

CHAPTER 12 FAIRNESS: SOCIAL   187
Intellectual Property and the Society   187
Environmental Issues   188
Experts and Paternalism   190
Social Aspects of Employment   191
A Real-life Case: Environmental Cleanup—Problems with the Superfund   192
Notes 193
Problems 193

SUMMARY   201
Some Words of Caution   201

UNIT FOUR   ADVANCED TOPICS

CHAPTER 13 RESOURCE ALLOCATION   205
What Is Resource Allocation?   205

Allocation by Merit 207
Allocation by Social Worth 207
Allocation by Need 208
Allocation by Ability to Pay 209
Allocation by Equal or Random Assignment 209
Allocation by Similarity 210
How to Decide among Methods 210
    A Real-life Case: Ethical Issues in Affirmative Action 212
Note 213
Problems 213

CHAPTER 14 RISK 221
A Historical Perspective 221
Defining Safety and Risk 222
Evaluating Risk 223
Making Decisions about Risk 225
Some General Guidelines 226
    A Real-life Case: Experimental Drug Testing in
    Humans 227
Notes 228
Problems 229

CHAPTER 15 DEALING WITH DIFFERING ETHICAL SYSTEMS 236
Differing Anthropologies 236
Differing Principles and Methods 237
Monism and Relativism 239
Postmodernism 240
True Pluralism 241
Conclusion 242
    A Real-life Case: Geological Experiments in Sacred
    Mountains 242
Notes 243
Problems 244

CHAPTER 16 HABIT AND INTUITION 253
Rationalist Approaches to Moral Action 253
Advantages of Rationalist Approaches 254
Problems with Rationalist Approaches 255
Toward a More Comprehensive Approach
to Moral Behavior 257
    A Real-life Case: The Ethics of Human Cloning 257
Notes 259

SUMMARY   261
Some Words of Caution   262

Cases: The Rest of the Story   *263*
Index   *265*

PREFACE

Purpose

Ethics in science and engineering has attracted increasing attention in recent years. Several well-publicized incidents, like the destruction of the space shuttle *Challenger* and the accusations surrounding the Thereza Imanishi–Kari/David Baltimore case, have focused heightened attention on the values by which scientists and engineers govern their professional behavior. In response, legislatures and governmental agencies have imposed ever more strict regulations regarding public disclosure, conflict of interest, and the like. Universities and national accrediting agencies are beginning to insist on formal training in ethics. Indeed, the Accreditation Board for Engineering and Technology (ABET) now specifically requires training in ethics for all engineering undergraduates. The National Institutes of Health requires formal ethics education for the graduate students funded by the NIH Training Grant Program.

No doubt this focus on ethics in science and engineering fits into a broader societal debate about personal and social morals in general. This debate is strongly colored by what many perceive to be a steady erosion of moral standards throughout much of Western culture. Regardless of whether such a decline truly exists, the perception of decline has made the debate about moral values increasingly shrill and bitter. Unfortunately, many scientists and engineers remain inadequately prepared to contribute to moral debates in a useful way, even within their own disciplines. Good intentions alone do not substitute for a keen eye for detecting ethical issues and a sound method for reasoning about them. This book seeks to remedy the problem, at least in part.

Unique Aspects

This book represents a new approach to education in technical ethics. The following describes briefly the unique aspects.

Broadened Intended Audience

This book has been written to appeal to both scientists and engineers. Indeed, most ethical issues facing scientists and engineers originate from the same problems. Conflict of interest, employer-employee relations, environmental awareness, and the like affect all technically trained people. This book focuses on issues that are common to all scientists and engineers, while explicitly highlighting the few differences where necessary.

Usability in Crowded Technical Curricula

Many colleges and universities offer a technical ethics course as an elective for three credit hours. Such courses reach relatively few students. Technical curricula are already too crowded to make a three-hour ethics course mandatory, so other ways are needed to meet the new requirements described above. Many ethics texts do not adapt easily to formats other than the traditional three-hour course. This book can be used in two different ways to solve this problem. The text in its entirety provides the basis for a one-credit-hour semester course, with each chapter corresponding to one lecture. Alternatively, each major unit forms a self-contained module of three to four lectures that can be dropped into any technical course with ease. Exposure to the entire text can then take place over four years, with Unit 1 in the freshman year, Unit 2 in the sophomore year, and so on.

Use of an "Ethical Serial" in Settings Familiar to Students

This text attempts to breathe new life into the study of ethics by employing an entirely new teaching device we call an "ethical serial." As with a television serial, each fictional case can be understood on a stand-alone basis. However, the cases use a single set of characters who develop their personalities throughout the book. Thus, the characters mimic real people far more closely than those in other ethics texts. Furthermore, the text breaks new ground by setting the fictional cases in situations directly familiar to college-aged students. The characters are students in science or engineering who mow lawns, baby-sit children, and fall in love while facing ethical issues identical to those of a professional environment. Finally, the cases employ real dialog instead of bland third-person narration.

Use of Virtue Theory

There are probably as many theories about ethics as there are ethicists. Surprisingly, however, the simple and time-honored approach based on virtues has made little appearance in texts for science or engineering. Virtue theory and science share a common cradle with the ancient Greeks, and involve very similar ways of thinking. Virtue theory seemed to pass out of favor among many ethical writers early in the twentieth century, reappearing with renewed vigor only in the mid-1980s. This approach provides a simple and easily remembered way for thinking about ethics that is ideal for a first course.

Innovations in Ethical Method

The technical disciplines pose many complex ethical questions. Unfortunately, for most scientists and engineers, their undergraduate training contains all the formal ethical training they will ever get. Thus, even in an elementary treatment it seems prudent to offer some approach to handling ethical complexity. Unfortunately, we believe that no current ethical theory (including standard virtue theory) adequately handles ethical complexity in a coherent, consistent, and understandable way. We recognize that many keen minds have applied themselves to this problem even at the advanced level. Nevertheless, we respectfully offer our own approach, mainly in Unit 2, that we believe is straightforward and practical yet comprehensive. This approach builds upon virtue theory, although we indicate clearly where we step beyond standard treatments. Our additions begin with the use of "event trees" to list systematically the consequences that follow from an action. (However, we avoid slipping into utilitarianism and related approaches by evaluating consequences in terms of how they square with the classical virtues.) Furthermore, we include the likelihood of consequences as part of the analysis in a way that has no counterpart in standard virtue theory. Finally, we take the ideas of intention and habit that are central to virtue theory and reframe them in terms of "interior consequences," leading to a more precise way to include these crucial ideas into ethical analysis.

Organization

We have organized the book into four units with four chapters each. Loosely speaking, the first half of the book focuses more on ethical reasoning, while the second half focuses on applications. The chapters in the first half of the book are largely stand-alone entities (except for Chapters 6 and 7, which depend heavily on each other) but are fairly sequential in nature (except for Chapter 8, which can be placed nearly anywhere). The chapters in the second half of the book are more completely stand-alone and do not need to be treated in sequence.

Unit 1 discusses the importance and scope of ethics, and offers a largely standard treatment of basic virtue theory within the context of science and engineering. Unit 2 discusses ethical complexity and moral responsibility within the same context. Unit 3 focuses on applying the virtue of justice, permuting the truth and fairness aspects of justice with the personal and social levels of human interaction. Unit 4 contains an admixture of more advanced topics, ranging from difficult applications (like risk and resource allocation) to deeper treatments of conflicting ethical methods and habit/intuition in ethical decision-making.

Each chapter contains one real-life case in ethics together with a few discussion questions. Because real-life situations tend to have many facets, these cases can serve as the basis for treatments varying from a brief class discussion to a full-length term paper. Each chapter at its end also contains several fictional cases that deal with the subject of the chapter and are more focused in their scope than the real-life cases. These cases should be suitable for class discussion or for homework problems of modest length.

Acknowledgments

First and most important, EGS thanks Juliann Seebauer, his loving wife, for graciously assuming extra duties in watching their two young children during those hours on Saturdays when he was tapping away furiously on the word processor. Both authors extend special thanks to Stacey Reiter, who helped greatly with editing early versions of the complete manuscript, and to Brian Meis and Juliann Seebauer, who helped to edit the fictional cases. We thank Donna Sarver, who helped to put the instruction manual accompanying the book in final form, and Eric Blomiley, who advised us in constructing the event trees in that manual. Bill Hammack has our gratitude for helping us test the questions in the virtue theory demonstration of Chapter 16. Finally, we thank the Department of Chemical Engineering at the University of Illinois/Urbana-Champaign for providing both tangible (i.e., monetary) and intangible support during crucial phases of this project.

UNIT ONE

FOUNDATIONAL PRINCIPLES

"The unexamined life is not worth living."
SOCRATES (C. 470–399 B.C.), QUOTED IN PLATO'S *APOLOGY*, 38

"The happy life is thought to be virtuous; now a virtuous life
requires exertion, and does not consist in amusement."
ARISTOTLE (384–322 B.C.), *NICOMACHEAN ETHICS*, BOOK X, CH. 6, 1177A

1

APPROACHING THE SUBJECT OF ETHICS

"Virtue is its own reward."

MARCUS TULLIUS CICERO (106–143 B.C.), *DE FINIBUS*

An Example

Consider the following fictional situation:

CASE 1.0 Truth in Writing a Resume

"Martin, can you take a look at this for me?" asked Myra Weltschmerz as she handed a copy of her resume to her boyfriend Martin Diesirae. "I want to turn this in to the engineering placement office tomorrow."

Martin sat back in his chair scanning the document while Myra stood waiting in front of him. Both were seniors at Penseroso University, he in computer science and she in environmental engineering. They had been together for two years. After a few moments, he raised his eyebrows and declared, "I don't know how you can put some of this stuff on here. You're basically lying!"

Myra cowered noticeably. "Martin, what do you mean? I'm not a liar."

"Come on! You are so! Look at this, under 'Work Experience.'" He leaned forward and held the paper about 6 inches from her face. "It says 'Accountancy Consultant to Baxter Brothers, Bakers.' That's garbage!"

"But I was—,"sputtered Myra, drawing back.

"You were nothing!" Martin broke in. "Mr. Baxter was your next door neighbor. All you did was come in once on a lark and teach his daughter how to use a spreadsheet. She was a part-time clerk! Then she entered stuff she got from the real accountant. You didn't even get paid!"

"Y-yes I did," stammered Myra timidly. "Mr. Baxter slipped me a twenty and told me and Karen to use it on a pizza."

"That's pathetic," Martin snorted. "No recruiter will take it seriously. Or what about this, under 'Extracurricular Activities'? You put 'Dixieland Jazz Band Ensemble.'" Martin jabbed his finger into the air at her. "You went to one meeting before you dropped out."

"But I paid the dues," Myra offered lamely.

"Hah! I could pad my resume like crazy just by paying dues," Martin snapped.

"But I'm a senior looking for a permanent job. A lot of other students have stuff like this on their resumes. The woman at the placement office said I should try to make my background look special."

"Your background isn't special. That's just the breaks," sniffed Martin. "I've told you for a long time to quit all your baby-sitting and do something more impressive. You must spend eight hours a week for that woman . . . Dolores or whatever. And she doesn't even pay you that much because she's always broke." Martin threw up his hands. "But you never listen."

"But Dolores needs help, and I like children. That should count for something."

"Hey, it doesn't count for much on a resume. You don't even have it on here!" He tossed the paper on the desk. "You can be so stupid, sometimes."

- Is Myra's representation of the consultancy acceptable? Why or why not?
- If not, what should she do?
- Is Myra's representation of the jazz band membership acceptable? Why or why not?
- If not, what should she do?
- Did you arrive at your answers immediately, or did you need to think for a while?
- Do you think most people would recommend what you did?

Let's consider in more detail how to approach questions like these.

The Importance of Ethics in Science and Engineering

Broadly speaking, scientists seek a systematic understanding of the physical world. Engineers seek to apply that knowledge for the practical benefit of all people. Most students in these disciplines will readily agree that mastering them requires long hours of grueling effort. Nevertheless, the effort seems worthwhile not only because success can offer a decent living, but also because the fruits of this work influence life in every corner of the planet. This wide sphere of influence makes working in the technical disciplines very exciting, but should also give us pause. As soon as what we do in our professional lives affects other people, our ethical judgment comes into play as well as our technical judgment. There are three good reasons we should give as much attention to developing our ethical skills as our technical ones.

First, good ethical behavior usually leads to good consequences, both for ourselves and for society at large. Sometimes the good effects show up immediately, as with a reward for returning a lost wallet. Other times the effects come much later, as with trust and respect from our colleagues. Some might argue that unethical behavior sometimes pays big dividends, as with stealing secretly from a cash register. However, the long and bloody trail of human history, running from the wholesale slaughter of the Dark Ages to the recent warfare in Kosovo, suggests that injustice leads mainly to suffering in the end.

Second, scientists and engineers make decisions crucial to society at large, and therefore shoulder an enormous burden of public trust. The complexity of modern technology forces those untrained in its way to depend on scientists and engineers for expert judgments. Unfortunately, the increased specialization of scientists and engineers sometimes leads to a narrow focus that cripples their ability to make and explain these judgments. This handicap carries over into ethics. When important and complex questions of right and wrong confront scientists and engineers in their professional work, they sometimes find themselves inadequately prepared about how to approach the issues or to communicate their advice clearly. Formal study of ethics can help to overcome these problems.

Third, happiness comes from reasoning through a complex moral puzzle, choosing a good course of action, and following through. Of course, people can sometimes do what is good on the basis of gut instinct alone. As thinking beings, however, people tend to find more satisfaction in understanding why they do what they do. Indeed, over two millennia ago Aristotle identified good ethical thought and action as the ultimate source of human happiness.

Unfortunately, education in science and engineering often provides little guidance in how to think about right and wrong. Our society at large knows this, and is sometimes uncomfortably willing to accept the movie clichés of "mad scientists" or engineers who act as unwitting pawns of larger evil forces. Granted, almost all of us receive a great deal of moral training from our parents. Primary and secondary education adds its contribution, and formal religion offers even more to its believers. However, the work place in science and engineering presents a distinct set of ethical problems. These problems often prove quite complex, and we need approaches that rely on more than gut instinct or simple rules learned in childhood. This book attempts to fill the gap in part by introducing the study of ethics applied to science and engineering.

Managing Ethical Discussion

Discussions of what is right or wrong, good or bad, often leave some people feeling ill at ease. There are several reasons.

First, how can we avoid name-calling and bruised egos in ethical discussions? It helps to distinguish between what a person says or does, and who that person is. Each of us represents some mixture of good and bad. Good people sometimes do bad things and vice versa. In other words, the goodness of a particular act or attitude does not determine the ultimate goodness of the individual. Furthermore, growth in the moral life takes time. Some people progress faster than others, at rates that depend not only on personal effort but also on all sorts of uncontrollable environmental factors. We cannot justly criticize someone for being molded in part by forces of culture and upbringing.

Second, how do we deal with the ambiguous, hard-to-define concepts that lie at the heart of ethics? Scientists and engineers, whose training normally deals with precise mathematical relations and sharply defined categories, sometimes experience frustration with reasoning qualitatively. Some may even dismiss the effort as meant for softer minds that can't handle complicated subjects like differential equations, thermodynamics, or quantum theory. This viewpoint ignores the fact that interpersonal relations, management, policy-making, and sales require far more skill in qualitative thinking than in quantitative. Unfortunately, some of the words used in moral discussion do carry many shades of meaning.[1] When unrecognized, such differences in usage often lead to irreconcilable disagreement. However, careful attention to exactly how words are used can help to avoid such problems.

Third, how do we deal with unpleasant memories of earlier wrongdoing? That all depends on what kind of people we hope to be. Errors and mistakes are part of human life. If we hope to grow into wiser people, mistakes can teach us what to avoid. Temporary guilt feelings help to burn these lessons into our minds in the way a hot iron brands a cow. However, there is no point in letting the brand burn for too long. Guilt that refuses to resolve itself becomes destructive and paralyzing, and usually points to deeper parts of the emotional life that need attention. A willingness to accept hard lessons combined with a commitment to continuing improvement can help us avoid falling into a rut.

Philosophy, Religion, and Ethics

Who should pronounce final judgment on right and wrong? Over many millennia people have appealed to judges, kings, and religious leaders for such judgments. The disappointing result has often been grand declarations claiming complete knowledge and eternal truth. History, of course, has usually deflated these claims. Some moral questions seem unanswerable on a purely natural level—that is, a level that appeals only to what people can observe and test in the physical universe. To proceed further seems to require an appeal to the "supernatural" level—that is, a level outside the observable physical universe. Such supernatural appeals have played such an important role in moral thought that we must decide right at the outset how to handle them.

In fact, many systems of thought and action have spoken to questions of morality over the centuries. We can loosely classify these systems as either "philosophy" or "religion." Since disagreements and misunderstandings sometimes arise over what these words actually mean, it seems prudent to offer brief (though incomplete) definitions here:

Philosophy: the rational study of principles governing knowledge, conduct, and the nature of existence.

Religion: a set of beliefs and practices concerning the supernatural, conduct, and the nature of existence. Religion appeals to one or more superhuman beings as governing forces for the physical universe.

Religion differs from philosophy by referring to supernatural beings and to things that must be taken on faith. Philosophy customarily avoids such references. Also, religion prescribes specific practices designed to promote good moral conduct, and may include paradoxes that confound reason. Philosophy, on the other hand, demands no devotional or ritual observances, and lies purely in the realm of reason.

Despite these differences, both philosophy and religion say things about moral conduct based on reason or faith. Not surprisingly, systems of thought and action that appeal to the nonphysical world cannot be checked by systematic experiments. Thus, many philosophies and religions coexist, with no agreement on how to pick the "correct" one, assuming a "correct" one exists. Herein lies an unsolvable problem for ethics. Each system depends upon different ideas about human existence, which in turn lead to significant differences in moral rules.

This book cannot settle such differences. Its description of human existence remains at a purely natural level, staying away from supernatural concepts like "revelation" and "god." Unfortunately, as we have said this perspective proves inadequate for tackling certain ethical problems; we need additional principles. These principles resemble the axioms used in mathematics. For example, classical geometry relies upon certain axioms about how line segments and angles add together, how parallel lines relate to each other, and so on. Given these axioms, we can derive all kinds of consequent theorems (to the agony of many high school students!) that compose the main body of classical geometry. Other axioms lead to other kinds of geometries.

Similarly, this book invokes a small number of axiomatic ethical principles as they are needed. These principles originated with the philosophers of ancient Greece and are shared by most Western religions and philosophies today. For the Greek thinkers, science formed merely one aspect of a much larger philosophy that also dealt with morality. Aristotle, who lived over 2300 years ago, drew distinctions between science and other branches of philosophy. His thought still maintained a close connection between them, however, as with his development of formal logic. The deep split that developed between experimental science and speculative philosophy originated much later with the Enlightenment of the eighteenth century. That split still exists today. Nevertheless, modern science and the ethical principles asserted here share a commonality that traces back to the cradle of Western civilization. Thus, these principles will not seem surprising, particularly to most scientists and engineers.

In short, this book uses a self-consistent world view that is compatible with both modern scientific thought and most Western philosophy and religion. However, the fundamental principles asserted in this book stand only upon their intuitive reasonableness and their long tradition of use. Further justification requires an appeal to something beyond the observable world. This book makes no such appeals, but points out explicitly where they would prove helpful.

The Existence of Right and Wrong

Interestingly, we must begin our study of ethics by adopting an axiom regarding the most fundamental question one can ask about right and wrong: do they exist in any objective way? Some people argue that all truth is little more than personal opinion—that culture and upbringing completely bias any ultimate judgment. This book avoids such extreme relativism. We will instead adopt a view that meshes better with science and engineering. A scientist or engineer takes for granted that certain laws of physics, such as $E = mc^2$ and $F = ma$, operate under all circumstances. We will assert a related ethical principle. Since we make such assertions so rarely in this book, we will highlight them as they appear.

Principle: *Certain aspects of right and wrong exist objectively, independent of culture or personal opinion.*

This principle does not declare exactly which things in ethics exist objectively, but despite its imprecision the statement still finds its strongest defense through philosophy or religion. This principle has the important practical consequence of moving ethics closer to discerning an objective reality rather than defining a subjective standard.

The Subject of Moral Analysis

Having proposed that objective morality exists, we might ask which matters lie within the moral domain and which do not. *In classical moral thought, morality concerns the goodness of voluntary human conduct that affects the self or other living beings.* Let's look more closely at what this definition really means.

First, the word "voluntary" holds great importance, implying that we have adequate control over what we're doing. *Assuming we have not deliberately allowed ourselves to remain ignorant, powerless, or indifferent, we have complete moral responsibility for what we do only with adequate knowledge, freedom, and approval.* It seems both unfair and imprudent to hold people responsible for meeting a standard of behavior they cannot reach because of normal human limitations.

Second, the definition restricts the object of moral behavior to living things. That is, you cannot behave morally toward a rock, except when that behavior indirectly affects some other living thing (like throwing the rock at your next-door neighbor).

Third, the definition uses the word "moral" rather than "ethical." What is the difference? In fact, the two overlap heavily. "Moral" generally refers to any aspect of human action. "Ethical," on the other hand, commonly refers only to professional behavior. Since this book concerns itself principally with situations encountered in professional life, "moral" and "ethical" will often appear interchangeably.

The Role of Codes of Ethics

Many professional and scholarly societies maintain formal codes of ethics. Such codes seem to find more use in engineering than in science, probably because engineers tend more often to view themselves as members of a profession like medicine or law. Such codes remind society members of the high ethical standards expected in the work place. Also, codes lay out those standards to new workers who have little experience. Finally, as public documents, codes can help professional societies take formal or legal disciplinary action against flagrant violators.

However, codes suffer from severe limitations in the rough-and-tumble of the real world. Codes lay out general ideals of ethical behavior, and often establish specific rules for commonly encountered situations. However, no list of ideals and rules can possibly give adequate guidance in all the complex situations that can arise. Shades of gray abound, and the best way to apply ideals and norms may not be obvious. Moreover, focusing only on the specific rules in codes sometimes leads to ethical minimalism, which is the idea: "If it's not specifically forbidden, it must be allowed." In addition, some situations call for quick decisions, with no time to consult a "rule book" of any sort. Worse yet, often no "traffic cop" is around to blow the whistle on code violations. Finally, certain formal ethical standards can change with time, sometimes in response to legal decisions.

All these shortcomings point to a need to develop ethics that spring habitually from the inside, and do not depend on some external list of rules. Strong ethical character makes it easier to rapidly and consistently handle messy situations not listed in a code.

A REAL-LIFE CASE: Destruction of the Spaceship *Challenger*

Shortly before noon on January 28, 1986, the U.S. space shuttle *Challenger* lifted off from its launching pad at Cape Canaveral, carrying several astronauts and a schoolteacher. Seventy-two seconds later the spaceship disintegrated in a fireball. A subsequent investigation showed that cold temperatures on the morning of the launch reduced the resiliency of the O-rings that sealed joints in the solid rocket boosters. Both the primary and secondary O-rings failed to make sealing contacts, permitting hot exhaust gases to escape and penetrate the adjoining fuel tank filled with liquid hydrogen and oxygen.

The problem proved to be no surprise to the booster manufacturer, Morton Thiokol. Indeed, engineer Roger Boisjoly had completed bench tests nearly a year earlier showing that O-ring sealing properties were lost for several minutes below 50 degrees Fahrenheit. However, under pressure from Congress to keep costs down and an aggressive launch schedule intact, neither Thiokol management nor NASA officials showed any interest in redesigning the joint. Because of the 18-degree temperature on the night preceding the launch, Boisjoly and other Thiokol engineers recommended strongly that the launch of January 28 be aborted. However, this recommendation was overruled by Thiokol management and NASA.

Clearly this case illustrates some serious lapses in judgment. The seals have since been redesigned. Nevertheless, current estimates of the chance that a given shuttle launch will fail catastrophically from some cause lie at 1 in 248. Given the large number of shuttle launches anticipated for scientific purposes and for construction of the new space station *Freedom*, the cumulative probability of disaster becomes significant. Furthermore, observers point out that some engineers at NASA have become so obsessed with avoiding blame for future trouble that they demand endless reports and studies that actually wind up increasing risk.

◆ How safe should the shuttle be before it is allowed to fly?

◆ What kind of management system might avoid both carelessness and paralysis?

References

Kiernan, V. "Safer Shuttle Still Risks Catastrophe." *New Scientist* 6 (1995):145–151.

Vaughan, Diane. *The Challenger Launch Decision.* Chicago: University of Chicago Press, 1996.

"It is proof of a base and low mind for one to wish to think with the masses or majority, simply because the majority is the majority. Truth does not change because it is, or is not, believed by a majority of the people."

GIORDANO BRUNO (1548–BURNED AT STAKE 1600)

Note

1. Take the word "good," for example. Suppose you heard someone say, "John deserves a real pat on the back—he stood his ground in the face of bitter opposition and did some good!" What image springs to your mind in response to this compliment? Maybe John is a saint-in-waiting fighting for the downtrodden. On the other hand, maybe John has merely argued for an attractive color of paint on the office walls. Who can argue that aesthetics is not a "good" of sorts? So "good" may refer to nonmoral as well as moral considerations.

Problems

1. Write a page or two describing an ethical dilemma you have encountered in a job you've had. (If you've been lucky enough never to have been confronted with a problem like this, describe one that a friend or relative of yours has had.) Recommend what action you think you (or your friend/relative) should have taken, and give reasons for and against that recommendation. Note: you don't have to say what was actually done in real life (unless you want to)!

2. Each case below has a question after it.

   a. Begin to put together your answer by writing down a brief list of options available to the main character who has to make a decision.

   b. Under each option, write a bulleted list of reasons for and against that course of action. The reasons should be short—no more than a phrase or sentence per point.

   c. Recommend what you think the character should do.

CASE 1.1 Endorsements and Commercialism

Myra Weltschmerz and her boyfriend Martin Diesirae walked slowly hand in hand outside the Engineering Library at Penseroso University. They had been studying hard together, so the warm twilight of early autumn offered an inviting break. Penseroso stood near the center of Exodus, a midwestern city of about three hundred thousand. Even so, the air carried the sweet rural scent of the grain harvest. Venus and Jupiter shone brightly as evening stars against the pastel sky. Myra squeezed Martin's hand. She had lived a hard life, but for an instant all seemed well with the world.

"You know," she whispered, "I had a good interview today."

"Good!" he responded. "Did you get your resume fixed up like I told you?"

"Not exactly. But I guess you did have some good points. I took out that bit about consulting, but I left in the Dixieland Jazz Ensemble as an extracurricular activity. I also put in my baby-sitting for Dolores' kids as 'child care.'"

Martin nodded. "It's a step in the right direction. I still think the jazz band is a stretch, but the 'child care' has a good ring." He paused, then continued wistfully, "Yeah, it's all about sales. You just have to sell yourself right." Then he grew earnest. "Did I tell you? Paragon Academic Supply called me today. You know that place a couple of blocks from here? Somehow they got wind that I won the Computer Science Department's Byte Award this year for scholarship and activities. They're starting some new promotion in the next month, and want to feature local students using their products."

Myra stopped in her tracks. "Really?!"

"Yeah. Awesome, huh? Anyway, they want to take my picture for a poster and newspaper ads, and get some quotes about how great I think their store is. In return, they're offering me a certificate for five hundred bucks of their merchandise. Retail value, of course." Myra stared for a moment, not knowing what to say. "That's a lot of stuff," Martin continued. "I could put it toward a new printer for my computer, or just get a fancy new calculator. I guess I don't have to decide now. Anyway, they asked me to sign the contract tomorrow."

"So they're going to give you this just because you won that award?" Myra asked.

"Yeah. Is something wrong with that?"

"I'm not sure. Pro sports figures sell their names all the time. And so do other famous people. But I've never heard of a student doing it. I mean, being a good student means learning a lot, right? It's not like you're doing something for someone else. Learning is supposed to help you all by itself. Do you really need to get paid extra if someone thinks you do it well?"

Martin stiffened. "Obviously you're not too excited about this!" he exclaimed. He pulled his hand away. "I thought you'd be happy."

"Martin, that's not what I meant! It's just that everything is so commercial these days already. Does studying have to get commercialized too?"

"I don't see the problem," Martin huffed. "I do a good job, get some recognition, and use it in a perfectly legal way." He eyed her suspiciously for a moment. "You're not jealous, are you?"

Myra felt her stomach tense. "No, Martin, I'm happy. I know it's very important to you." She paused, and her face hardened slightly. "But . . . are you going to look for an agent every time you get an A on a test?"

Martin threw up his hands. "I don't understand you! You say you're happy, but you don't show it. Do you want me not to sign?"

Myra's face grew pale. "I don't know what I want."

"You can say that again! That's always been your problem."

"Martin!" she whined.

◆ Should Martin accept the endorsement contract?

CASE 1.2 Reporting Apparent Bribery

Mind adrift, Celia Peccavi shuffled slowly from the restroom of Pandarus Pizza to the kitchen. She had worked at Pandarus for six months, and despised slow days like this one. The minutes turned to centuries. She would quit in an instant, but as a sophomore in biology at Nosce te Ipsum University, Celia desperately needed the money to pay her tuition.

As she drifted past the manager's office, her ears perked up at the sound of animated conversation inside. She did not recognize the voice. Celia never missed a chance to eavesdrop, so she lingered just outside the closed door, where with effort she could make out the words.

"Mr. Mauvais, I really need this job," came the unfamiliar voice. "I can paint your windows better than the person you have now. I've helped several restaurants with their promotions. You should see my work!"

Celia heard the squeak of an office chair, then the voice of her manager Thorne Mauvais. "I don't know. I'm happy with the artist we have. His work is good and his rates are fair. He's very reliable. I'm not a public works agency, you know."

"But my work is better. What is he charging you?"

A pause ensued, and Celia heard some papers shuffle. Apparently Thorne was showing some figures to the visitor, who moaned, "I can't beat that price! That's not much over break-even! Are you sure that's right?"

"I told you his rates were fair," retorted Thorne.

Another pause followed. "Look," the visitor broke in suddenly. "I can't take the job this cheap. But how about this? I'll charge 10 percent over what Pandarus is paying now, but I'll throw in something extra for you personally. You said earlier you like football, right? Well, I have a friend who has season tickets for the Penseroso Peacocks. Fifty-yard line, close to the field. He lets me use them all the time as a favor. I'll get you tickets for two for any game you want this season."

"Any game?" Thorne repeated.

"Any game!"

"Well," Thorne murmured vaguely. "I don't know. The Peacocks aren't too good this year." He stopped, then continued, "Toss in a second game and you've got a deal."

"Two games? Two tickets each?"

"Yup."

The visitor sighed unhappily. "OK, I'll do it."

"I'll call you tomorrow with the details of when you should start with the window painting, and with the games I want."

Celia heard the two rise out of their chairs. She started instantly toward the kitchen, but the office door opened before she got more than a few steps. "I look forward to hearing from you, Mr. Mauvais."

"Sure," Thorne responded patronizingly.

Celia turned around involuntarily to see them. Her glance met Thorne's, and his face darkened. "Celia, come into my office, please," he called, beckoning.

She complied. When they got inside, Thorne took a seat and motioned for Celia to shut the door. "Why aren't you in the kitchen?" he inquired suspiciously.

She shrugged. "I had to go to the bathroom."

Thorne's voice grew accusing. "The bathroom is right next door, but I haven't heard anyone in there for a while. You weren't just hanging around, were you?"

Celia tossed her long brown hair. "Hey, I had to fix my hair. It needs a lot of work sometimes. I had to brush it out. Do you mind?"

"This isn't a beauty salon. I don't want you primping during working hours." He pointed at her hair. "Anyway, it doesn't look any better than it did earlier," he said derisively.

Celia stiffened, then counterattacked. "So what did that guy want?"

"None of your business! Get back to work!" Thorne shot back.

With minced steps, Celia opened the door and sauntered out, tossing her hair with a flourish. "Maybe we should ask the owner about company policy toward visitors," she remarked sarcastically.

"Never mind the owner. I run this place!" Thorne yelled after her.

◆ Should Celia tell the owner about what she heard?

CASE 1.3 Obeying a Law "for the Sake of It"

Terence Nonliquet stretched slowly in the passenger seat of the cramped car. It was Friday night, and his head ached from a long week of studies and duties as a teaching assistant. He regretted agreeing to accompany his girlfriend Leah Nonlibet to visit her ailing grandmother. The old woman's life was slowly ebbing away, and Leah became very depressed after each visit. This time she asked Terence to come along as a support, and to help with the four hours of lonely driving. Terence knitted his brow as he dreamed about all the other ways he could be spending the evening. "How much further?" he asked with a sigh.

"About 80 miles," Leah responded. "It's an easy drive now. It's a straight shot from here on this two-lane road."

"Good. It's dark and there's nothing to see."

Leah tried to change the subject. "How was being a teaching assistant this week?"

Terence furrowed his brow. "Pretty hard, but I guess I expected that for my first semester teaching. You know, I'm still only a junior. When the Computer Science Department has to use upperclassmen as teaching assistants, they usually go for seniors. But early this week I overheard one of the professors say my record was good, and they were really shorthanded for some reason."

During the conversation, Leah had braked to a stop for a red light at a lonely intersection. The light had been red for about a minute. "I don't understand why this light is so long," Leah muttered. "Every time I take this road, I get caught here for a couple of minutes. There are hardly ever any cars."

"I don't see any cars now," Terence rejoined, "and the place is flat, so you can see for miles. Why don't you just go?"

"Through a red light? I could get a ticket!"

"Do you see any cops?" asked Terence, looking around. "I don't. You're not going to get a ticket. Not this time."

"But how can you be sure? Maybe there's a cop car hiding where we can't see!"

Terence's voice sharpened. "Leah, just go. If you get caught, I'll pay the ticket. I'm wiped out from all this driving. Let's just move and get to your grandmother's."

"But I don't want a ticket on my record," Leah persisted defensively, "and maybe my insurance rates will go up. Anyway, it's just wrong to go against a red light. I won't do it."

"Why not?" Terence argued. "Stoplights are there to control traffic so accidents don't happen. You know there won't be an accident here."

"It's against the law!"

"The law is there just to prevent accidents. If you know there won't be an accident, the law isn't a big deal!"

"Yeah, right . . . what cop is going to buy that?" contended Leah testily. "The law doesn't say, 'stop at reds only when there's a chance of an accident.' It says, 'stop at reds,' period. I think it's always wrong to disobey traffic laws. You notice, I don't speed either."

"My aching butt is reminding me of that right now," Terence groaned. "You're more likely to have an accident from getting angry at me than from speeding or red lights!"

"I'm not angry."

Terence rolled his eyes and settled deep into his seat. Silence followed.

◆ What should Leah do?

CASE 1.4 Public Trust: The Duties of Club Officers

Todd Cuibono and Emily Laborvincet sat working on their homework together in his dorm room at Penseroso University. Their romantic involvement had begun a year ago when they met in a shared general chemistry class. Now that the fall semester had begun, both were sophomores: she in chemistry and he in chemical engineering.

The evening grew long, and Emily looked up wearily from her books, hoping for a break and a little companionship. "Hey Todd, I saw another one of those posters announcing a meeting of the student chapter of the National Chemical Society. I never heard of that—it's not the same as the regular professional group, the American Chemical Society, is it?"

Todd's face remained buried in his book. "No, it's different."

"So what is NCS?" Emily persisted.

Todd started to scribble in his lab book. "It runs sort of in parallel to the student chapter of ACS. In a lot of ways it does the same thing."

Emily remained puzzled. "Why have two groups do the same thing? It's stupid!"

"It's not stupid," Todd snapped. "In fact, I could become NCS vice-president next week if I accept the offer."

Emily's jaw slackened. "You never told me about this!"

"You never asked," retorted Todd. "The NCS president, Waldo Drake, asked me a couple of days ago."

"What do you mean 'asked'? Doesn't there need to be an election?"

"Not in NCS. It's still a new organization. It doesn't have a regular succession of officers yet. Waldo formed it last year after resigning as vice president of ACS. He had a big policy fight with the ACS president, Regina Livia—I forget why. Now his current NCS vice-president has graduated, and he needs a replacement."

Emily couldn't believe her ears. "Policy fight? I thought Regina was Waldo's girlfriend! So they broke up in a big way, huh?"

"Well, that too," replied Todd. "But their disagreement was mostly professional."

"I doubt that," laughed Emily. "I've noticed NCS events almost always conflict in time with ACS events. The two groups seem like rivals. Anyway, how does NCS get any money for its events? Despite the name, I bet they don't have a national organization to back them."

"Waldo has unbelievable connections," Todd responded. "And he knows how to use them. For example, his older brother graduated a year or two ago from Penseroso with straight A's, also in chemistry. He gave Waldo all his course notes, exams, and homework. They're awesome. But Waldo makes them available only to NCS members, and the member dues are really high. Still, a lot of students have coughed up the cash just to get at those files."

"Unbelievable!" Emily gasped.

"There's more. Waldo's younger sister is on the cheerleading squad for Penseroso. She's an absolute babe, and has several similarly gifted friends. Waldo has persuaded her and her friends to help out with car-washing fund raisers. They've done two or three so far. They use an off-campus lot owned by some other friend of his. The girls dress up in bikinis and sometimes heels, and the cars almost crash into each other trying to line up. The view also gets most of the NCS guys to help out."

Emily's eyes widened. "And you were there? You never told me!"

Todd shrugged. "Sure I was there. Why not? I mean, it was a public street corner after all . . . in full view of everyone. I didn't bother to invite you because you were busy every Saturday they had one.

"I think those car washes are disgusting!" Emily shot back. "I can't believe Waldo could persuade those girls to do that, even if one of them *is* his sister."

"Well, I guess he gives them a pretty big chunk of the proceeds. On the side of course," replied Todd.

"I can't believe that's legal!" cried Emily. "Don't the campus codes for organizations forbid that?"

"Actually, Waldo is following the codes to the letter," responded Todd. "The codes require that all money a registered organization takes in from on-campus activities be deposited in a special university account. All sorts of rules govern how that money can be spent. But regular dues and money from off-campus activities are exempt. The officers can use those funds basically at their own discretion. Anyway, NCS also gets money from the student organization fund. That's the money that pays for printing notices and the food at meetings on campus."

"But Waldo still has this slush fund on the side?" asked Emily. "And why would you want to be vice-president?"

"Waldo mainly uses the profits to get good food at meetings and pay for member parties," responded Todd. "I'm thinking of taking the job because it would look good on my resume. Also, Waldo will graduate at the end of the year. He'll take all his connections with him, so the organization will probably die. Then he'll have to do something with the war chest he's built. My guess is that he'll probably just divide it among the officers."

"That's larceny! You'll get in trouble!"

"I don't think so. The university doesn't know about the money, and even if it did, Waldo has followed the letter of the codes. If the NCS dies, the money has to go somewhere, and that's at officer discretion. Anyway, no one will complain, since the rank and file has no idea what's in the kitty."

"You shouldn't take that money!"

"OK, OK! I don't know for sure if there'll be any dividing of spoils anyway. But if it makes you feel better, I can reject my portion if I want. I'm still thinking of taking the vice-presidency, though, for my resume."

- ◆ Should Todd accept the vice-presidency?
- ◆ Should Todd accept any money?

2

THE PERSON AND THE VIRTUES

"To live in accordance with nature is to live in accordance with virtue."

<small>ZENO THE STOIC (C. 335–C. 263 B.C.), QUOTED BY PAUL MORE IN *HELLENISTIC PHILOSOPHIES*</small>

To rate the morality of an action, we must keep in mind how we are constructed as human beings. Put another way, we must describe an anthropology and understand how it influences human behavior. Only then can we begin to examine moral principles and methods. Accordingly, this chapter develops a simple but useful anthropology and examines how it affects moral behavior through habits called "virtues."

Developing a Model for the Person

Given the complexity of the human person, we cannot expect our anthropology to reproduce every aspect of how people behave. Instead, we must develop a simplified model, focusing entirely on morality. As with any model in science and engineering, we have first to decide to what degree of accuracy we want to represent our subject. Greater accuracy usually requires more complexity in the model. For example, at low speeds, classical Newtonian equations like $F = ma$ represent the motion of objects quite well. However, near the speed of light this model breaks down and must be generalized to include the complicated effects of special relativity. Does this breakdown mean that Newtonian mechanics is a bad model? Not at all! Newton's equations are fairly straightforward to use and provide results that are extremely accurate in most cases. Such features distinguish good models in general: good correspondence to reality under most circumstances with modest effort. Other examples of such models include the ideal gas law ($PV = nRT$) and the "lock-and-key" model for enzyme action.

This book offers merely an introduction to ethical thought, so it makes sense to employ a fairly simple model. Our model will focus only on moral behavior in typical adults. (Most scientists and engineers should qualify as "typical adults"!) We will not expect this model to make quantitative predictions, as might social scientists who perform statistical analyses of large populations.

Components of the Psyche

The anthropology we will employ was originally developed by the ancient Greeks. It views the person as composed of the senses and the psyche. The five senses of sight, hearing, taste, smell, and touch provide raw data about the outside world. The psyche puts these data together into a coherent perception and understanding that we commonly call "consciousness." As illustrated in Figure 2.1, the psyche comprises three parts:

Mind: The mind corresponds in many ways to a computer; both maintain memory and logic functions.[1] The mind classifies abstract concepts in a coherent way and uses them according to logical rules. In moral decision-making, the mind puts together sensory data from the present with memories from the past to predict what will happen in the future.

Emotions: The emotions are conscious, nonrational psychic responses to data from our senses and to certain kinds of internally driven neurochemistry (sickness, hormonal swings, various drugs, and the like). Emotions form the clearest link between the psyche and the body. Indeed, many emotions carry physical responses with them. For example, the heart beats faster with anger, and the face flushes with embarrassment. This connection explains why we often refer to emotions as "feelings."

Will: The will decides among alternatives presented to it by the mind in a way colored by the emotions. Exercising the will gen-

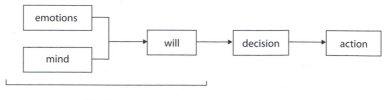

Psyche

FIGURE 2.1 A simple model for the origin of moral action

erally involves rational thought, and so at first glance the will might seem to be part of the mind. However, emotions make the decision-making process in humans very different from that in computers. Computers make decisions on the basis of cold logic. In humans the emotions also enter into play, not only coloring the process of deciding but also providing a crucial push to really act on the decision. Thus, most moral writers consider the will to be distinct from the mind.

In reality, it's difficult to dissect the psyche into distinct elements. All three flow into each other in the same way that the primary colors flow into each other to produce all the colors on a painting palette. The human brain is wired in a vastly more complex fashion than a digital computer, in which logic, memory, and decision functions can be readily identified and separated.

Limitations of the Model

Notice that this model says nothing about the unconscious mind, something for which modern psychology has accumulated overwhelming evidence. However, moral action requires action of the conscious will. The unconscious mind may push us toward certain kinds of behavior; psychological addiction, irrational compulsions, and explosive rages are obvious examples. Our model cannot account for all the behavioral problems that people have.

Notice also that this model says nothing about moral development. It leaves aside questions of how infants, who are not capable of moral behavior, develop into adults who are. Several developmental theories of psychology shed light on this subject, however.[2-4] Moreover, the model says little about the stages of moral development commonly observed even in adulthood. Once again, moral psychology and philosophy have much to say in this arena,[5-14] although the earliest theories for stages of interior growth date back nearly five centuries.[15]

Finally, this model says little about human community. A person is not an island; in fact, solitary confinement is a mode of punishment in many prisons. Communities from families to sports teams to church congregations to countries represent wholes that are greater than the sum of their parts. Thus, morality carries not merely a personal dimension but also a social dimension. Ethicists often discuss the social aspect in terms of "the common good," which simply means that the good of the community as well as the good of particular individuals enters into moral decision-making.

Habits and Morals

Our anthropology describes the person as a unity of mind, emotions, and will. How does this unity function morally? Many moral philosophers over several millennia have recognized the unique ability of the person to choose between good and bad. Just as important, these philosophers have recognized that this ability often does not involve a lengthy, drawn-out mental process for each choice. Rather, many simple moral choices occur with little thought because they have become habitual. That is, the mind, emotions, and will routinely combine forces to give nearly effortless moral action.

Of course, the goodness of that action depends on how the person has been trained. This training begins early in life, when for example toddlers are taught to avoid hurting others. Later, many children begin to regularly say "please" and "thank you." These actions do not come preprogrammed; they come from external training. As children grow older and become adults, they habituate themselves to behaviors of their own choice: holding doors open for companions, for example. These simple deeds smooth human interactions and are therefore good. While people not habituated to such actions may choose to do them at times, that practice often remains the exception rather than the rule.

Such habits govern more complex aspects of moral decision-making as well. In Aristotle's view, a class of good habits exists for each of the three parts of the psyche. Each class represents a *virtue*. Put another way, a virtue is the customary direction of one part of the psyche toward moral good. In simple situations, a virtue makes good moral action almost effortless. In more complex situations where the best choice may not be obvious, a virtue makes finding a good solution easier.

Mountain climbing represents a good analogy to using virtues. A mountain climber draws upon many simple skills that almost everyone has, like walking, pushing, and even breathing. However, reaching the top requires coordinating these skills precisely for the duration of a lengthy climb. Success demands that the muscles and lungs be well conditioned and practiced. Not surprisingly, a trained athlete reaches the top better than a couch potato of equal size and weight. The couch potato might be able to handle simple day-to-day living, but not the complex rigors of mountain climbing.

The Four Main Virtues

Because of the complicated interplay between the mind, will, and emotions, good actions depend upon all the virtues. Nevertheless, classical moral thought distinguishes four main virtues. Each one is rooted in a particular part of the psyche. These four typically bear the name "cardinal virtues" or "natural virtues." The discipline of theology has over the centuries defined additional virtues,[16] but we will not treat them here. The ancient Greeks observed that each of the natural virtues promotes behavior lying midway between excess and deficiency. Let's list what the virtues are:

Prudence: Prudence concerns the mind. A prudent mind thinks about a moral problem clearly and completely. Obviously, intelligence plays a role. However, the mind must also give itself enough time to work and must apply itself at the right level of detail. Hence, prudence involves not only intelligence but also forethought and practicality. The latter two distinguish this virtue as a mean. Forethought finds the balance between a snap decision and endless calculation, while practicality avoids both airy theorizing and drowning in details.

Temperance and Fortitude: Temperance and fortitude complement each other; both concern the emotions. In fact, these virtues deal with the opposite sides of emotional experience. Positive emotions (like happiness, affection, and amusement) draw us toward their cause, while negative ones (like fear, pain, and grief) push us away. Temperance controls our attraction, putting a brake on our impulse to move blindly toward something we like. As a mean, temperance avoids rashness, which gives in to a feeling completely. Temperance also avoids suppressing feelings altogether, which sometimes proves disastrous in the long run. Fortitude controls our aversion, putting a brake on our impulse to move blindly away from something we don't like. As a mean, fortitude falls between desperate evasion (again giving in to the feeling completely) and suppressing feelings altogether.

Justice: Justice concerns the will and has two aspects: truth and fairness. A will acting in truth chooses according to things as they actually are and not to illusions. A will acting with fairness seeks to give what is due to all concerned. Fairness distinguishes justice as a mean. In our relations with others, fairness sits between selfishness and complete neglect of self. In matters affecting others but not ourselves, fairness finds a balance among interests without showing bias toward anyone.

In describing the virtues, we need to keep a balance between their elements of choice and habit. Exercising the virtues resembles breathing the air; both processes can take place largely unconsciously. Nevertheless, breathing can come under conscious control if we choose. For example, weight lifters, swimmers, and mothers in labor consciously regulate their breathing. In the same way, humans often act habitually, but keep the ability to consciously regulate their moral behavior. That is, while we might customarily act with prudence, temperance, fortitude, and justice, we can also choose to do so.

Whether by choice or by habit, we should ultimately seek to do good things. How do the virtues fit into this goal? We invoke a principle that is important enough to highlight:

Principle: *People should always decide and act according to the virtues insofar as possible.*

This statement may seem obvious, but its complete defense probably needs philosophical or religious backing that we will not give here.

An Example

Let's look at a fictional case to observe the interplay of virtues.

CASE 2.0 Occult Compensation: Fairness to Employers

"I heard this play is pretty decent," Emily Laborvincet said to Todd Cuibono as they drove to the theater.

"Mmm," he responded as he glanced into the rearview mirror of his car.

She reached over and stroked his neck. "I'm glad you could get the night off work."

"Mmm," he said again. After a moment he continued, "I'm not losing that much anyway, the pay is so rotten. You know what? I've been asking around, and found that Pandarus Pizza pays a dollar less per hour than most of the other pizza places in town. And that factors in experience, type of work . . . all that stuff."

"So why not switch jobs?"

"The other places are further away from the dorm, and the hours don't fit my schedule as well. I checked."

"I guess life isn't fair," sighed Emily.

"You can say that again," he snorted. "But this time I'm evening the score."

"What do you mean?"

He strained to turn around while holding the car straight, and nodded toward the back seat. "You see that?"

Emily turned and saw a small bag full of breadsticks together with some desserts. All had come from Pandarus Pizza.

He continued, "I got that from Pandarus. Every two weeks I multiply a dollar by the number of hours I've worked. That's how much I figure I've been underpaid. Then I take food equal in cash value to that number. It's mostly stuff that we prepared but couldn't sell. Or sometimes it got old, or came from cartons that have been damaged at the restaurant, and aren't salable or returnable."

"Todd, that's stealing!" Emily gasped.

"No it's not, it's getting my fair due. The way I look at it, Pandarus has been ripping me off by not paying a fair wage. They can't use the stuff I take anyway."

"But . . . maybe some of it could go to a homeless shelter or something."

"Emily, I don't like to be called a thief," Todd snapped. "I didn't think this would bother you, or I wouldn't have told you. It's just like when I told you about taking the vice-presidency of the National Chemical Society, and you got upset. I finally turned it down to get you off my back. Let's not start all over again." He turned away and stared down the road in a way Emily knew well; further discussion would just spoil the evening she had looked forward to.

"He's getting really difficult in these conversations," she thought to herself. "I'm never going to bring this subject up again."

- Should Todd continue to take the food? Why or why not?
- Are there any situations in which the kind of thing Todd did could be justified?
- What should Emily do? Why?

The virtues (or lack of them) reveal themselves in several parts of this story. Consider first the pay scale of the restaurant. The virtue of justice requires fairness in the way employers pay their employees, so that $2.25 below the local average indeed raises suspicions of unfairness. Now consider what Todd was doing in response—he was stealing food to "make up" for the low wages. Secretly taking company property to make up for poor pay occurs commonly in the work place, so commonly that moralists give it a special name: "occult compensation." ("Occult" here means "hidden from view," not "mysterious" or "magical.") Occult compensation offends against the virtue of justice, both in fairness and truth. Occult compensation offends against fairness because an employee agrees voluntarily to work for specified wages, and fairness requires that the terms of the agreement be honored. With occult compensation, the employee in effect claims a larger wage without agreement of the employer. If the wages are unfair, the employee should try other ways to change them, like pushing for higher pay, participating in a strike, or ultimately quitting the job. Occult compensation offends against truth because the employee pretends that he or she is abiding by the agreement when the opposite is true.

What about the way Todd interacted with Emily in the car? Initially he practiced the virtue of justice by telling her the truth about what he was doing. However, she pointed out the injustice of his action. Todd then offended against temperance; he became too angry too quickly, and the conversation went downhill. However, he did finally catch himself by not letting the argument get completely out of control.

What about Emily's behavior? She acted with truth and fortitude by honestly calling Todd's behavior stealing, and by not backing away in the face of his protests. She acted prudently by following Todd's lead in not letting the argument escalate out of control. However, by resolving to never bring the subject up again, she acted contrary to fortitude. Fortitude requires that, in times and ways dictated by prudence, she continue to encourage and prod Todd to turn away from ways that are clearly wrong.

A REAL-LIFE CASE: Toxic Waste at Love Canal

In 1947, Hooker Chemicals and Plastics Company bought a parcel of land called Love Canal near Niagara Falls, New York, as a landfill for waste chemicals. By the time the landfill closed in the early 1950s, nearly 22,000 tons of wastes consisting of 248 different kinds of chemicals had been dumped there. Some of these wastes were exceedingly caustic, carcinogenic, or toxic, and included chlorobenzenes, polychlorinated biphenyls, and dioxin. During this time, Hooker did little to contain the waste, and contamination spread throughout the canal area. Children played in the polluted water, as no fences existed to keep them out. However, at the time relatively little was known about the harmful properties of many of the chemicals, and the local residents raised few complaints. Indeed, Hooker was a major employer in the area.

After the landfill closed, Hooker worked more diligently to contain the wastes. It lined the canal with impenetrable concrete, and placed a waterproof ceramic cap over the chemicals to prevent rainwater from entering. These precautions far exceeded common practice at the time. Shortly thereafter, the Niagara Falls Board of Education demanded the land to build an elementary school for the growing local population. In the face of eminent domain proceedings, Hooker reluctantly sold the land to the school board for one dollar, protesting that the land should not be excavated. Hooker inserted a clause into the contract absolving the company of any further liability for the land. However, the company did not disclose the details of what it had dumped or how much. For its part, the local government removed some of the ceramic cap when constructing the school, thereby permitting rainwater to seep into the canal. The government then sold some of the land for residential development, profiting handsomely. During the development, the concrete containment walls were breached several times with sewers and a storm drain.

Throughout the 1960s and early 1970s, there were scattered complaints from the residents about chemical odors. However, heavy rains during 1975 and 1976 raised the water table and caused some of the land to subside, creating cesspools of toxic, fuming wastewater that ultimately leaked into homeowner basements. Subsequent investigation revealed health problems that had developed among the residents over time, ranging from low birth weight to chromosomal damage. By 1978 the state of New York had relocated 238 residents, offering full financial compensation for their homes. A Presidential State of Emergency was declared for the area, and the federal government sued Hooker, by then a subsidiary of Occidental Chemical, for damages. In 1994, Occidental agreed to pay $120 million for a new containment facility plus continuing operation and maintenance costs. Litigation with the city did not finally end until 1998.

◆ Who was responsible for the environmental damage? To what degree?

◆ Which of the virtues were ignored, and by whom?

References

Deegan, John, Jr. "Looking Back at Love Canal." *Environmental Science and Technology* 21 (1987):328–331.

Hoffman, Andrew J. "An Uneasy Rebirth at Love Canal." *Environment* 37 (1995):4–9.

"Habits change into character."

OVID (43 B.C.–A.D. 18), *HERIODES*

Notes

1. As useful as this analogy might be, we should avoid carrying it too far, since computers possess no true consciousness or insight.

2. Jean Piaget, *The Origins of Intelligence in Children* (New York: Norton Library, 1963, originally published in 1936).

3. Jean Piaget, *The Moral Judgment of the Child* (New York: The Free Press, 1965, originally published in 1929).

4. Lawrence Kohlberg, "The Child as Moral Philosopher," *Psychology Today*, September 1968, 25–30; "Stages and Sequence: The Cognitive-Developmental Approach to Socialization," in *Handbook of Socialization Theory and Research*, D. A. Goslin, ed. (Chicago: Rand McNally, 1969).

5. Ibid.

6. Erik Erickson, "Eight Ages of Man," in *Childhood and Society*, 2nd ed. (New York: W.W. Norton, 1963).

7. For an incisive summary of these various developmental theories, see Daniel A Helminiak, *Spiritual Development: An Interdisciplinary Study* (Chicago: Loyola University Press, 1987), ch. 3.

8. Roger Gould, *Transformations: Growth and Change in Adult Life* (New York: Simon and Schuster, 1979).

9. Daniel Levinson, *The Seasons of a Man's Life* (New York: Knopf, 1978).

10. Lawrence Kohlberg, "The Implications of Moral Stages for Adult Education," *Religious Education* 72 (1977):183–201.

11. James Fowler, *Stages of Faith: The Psychology of Human Development and the Quest for Meaning* (San Francisco: Harper & Row, 1981).

12. Jane Loevinger, *Ego Development* (San Francisco: Jossey-Bass, 1977).

13. Carol Gilligan, *In a Different Voice: Psychological Theory and Women's Development* (Cambridge, Mass.: Harvard University Press, 1982).

14. Lawrence Kohlberg and Clark Power, "Moral Development, Religious Thinking, and the Question of a Seventh Stage," *The Philosophy of Moral Development: Moral Stages and the Idea of Justice, Essays on Moral Development*, Vol. 1 (San Francisco: Harper & Row, 1981), 311–372.

15. For example, the two Spanish Renaissance mystics John of the Cross and Teresa of Avila developed separate theories of development that are spiritual at their core but impact very directly on moral development. See St. John of the Cross, *The Ascent of Mt. Carmel*, in *The Collected Works of St. John of the Cross* (Washington, D.C.: ICS Publishers, 1979) and St. Teresa of Avila, *The Interior Castle* (Garden City, N.Y.: Image Books, 1961).

16. Examples include the "theological virtues" of faith, hope, and love. These theological virtues together with the natural virtues form the seven so-called classical virtues. Interestingly, there exists a corresponding list of seven "classical" vices: pride, lust, gluttony, envy, anger, greed and sloth!

Problems

1. Write a page or two describing an ethical dilemma you have encountered in some job you've had. (If you've been lucky enough never to have been confronted with a problem like this, describe one that a friend or relative of yours has had.) Recommend what action you think you (or your friend/relative) should have taken, and give reasons for and against that recommendation. Note: you don't have to say what was actually done in real life (unless you want to)!

2. Each case below has a question after it.

   a. Begin to put together your answer by writing down a brief list of options available to the main character who has to make a decision.

b. Under each option, write a bulleted list of reasons for and against that course of action. The reasons should be short—no more than a phrase or sentence per point.

c. Recommend what you think the character should do.

CASE 2.1 Accepting One Job, Then Another: Obligations to the First Employer

The clock struck 11 p.m. as Emily Laborvincet sat with her boyfriend Todd Cuibono in Frosty's 35-Flavors ice cream shop. Despite the late hour, Emily's gestures became increasingly animated as she poured out her concern to Todd, who by contrast sat impassively.

"I just don't know what to do," she observed wistfully. "I thought everything was settled when I accepted this work-study job for next semester at Ajax Foods. I sent them the acceptance letter last week. The people there seem pretty friendly, and the experience in food chemistry should be good. But now I have this fantastic offer from Tripos Metal Polish. It's a lot smaller than Ajax—only about sixty employees. But I'd get a lot more responsibility and independence."

"Sounds good," replied Todd.

"And I really like the owner, Monica Ichdien. There's something very special about her—she really seems to understand people. She said she could tell right off that I had a lot of potential, that I had something special. . . ."

Todd rolled his eyes. "One of those touchy-feely types, huh? Is she a psychic, too?"

"Todd, be serious. She's very practical when she needs to be. How else could she build a successful business? Anyway, she offered me a little part-time work this semester to get an early start and to let me earn some extra cash."

"So why don't you just take the Tripos job?" Todd retorted dryly.

"I don't think I would feel right. I voluntarily took the Ajax offer, and in writing! I think that's a promise of sorts, and I don't like to break my promises."

"But Ajax hasn't paid you a dime, and you haven't worked there yet," replied Todd. "It's not like they're paying you for work you haven't done. Anyway, it's September now; you won't work there until January. They have plenty of time to find someone else. If the job were that good, they probably had buckets of applications."

"I'm not so sure," she said. "A lot of the work-study students are firming up their plans right now. There probably *was* a long list to work at Ajax, but those students might have committed elsewhere by now. I just don't know."

Todd rolled his eyes again. "You worry too much. These companies don't really need work-study students. They just hire students to polish their image, and to get a leg up in recruiting for permanent jobs. Think of it this way. Suppose you take a job at some burger joint to earn some extra money. A week later, another job comes along at twice the salary with better hours. You'd quit the burger job in a nanosecond, wouldn't you?"

Emily looked perplexed and shrugged. "I don't think you understand. The job at Ajax is not a burger-flipping job. There's more permanence and commitment for both them and me."

"What, are you going to marry the manager or something?" Todd exclaimed.

"Todd, don't be so flippant. I'm serious. They were nice to me during recruiting, and promised to help tailor the job to my interests as much as they could. Still, it's true—the job will never give me the independence that Tripos would."

The edge in Todd's voice disappeared momentarily. "OK, you like the Tripos job better both for the money and the type of work. Why don't you push Ajax for more money? Maybe they'll do even better than Tripos if you threaten to switch jobs. Then the money will balance the type of work."

"But Todd, I already signed the acceptance. I can't renegotiate now!"

Todd's expression hardened. "Come off it, Emily. Anything is negotiable at any time. That's the way the world works. If Ajax wants you badly enough, they'll deal. If not, you walk away. It's as simple as that." He reached for the restaurant check. "It's getting late, and I'm tired."

◆ What should Emily do—take the Tripos job, renegotiate with Ajax, or let things remain as they are?

Case 2.2 Using Flextime Properly

Martin Diesirae's phone rang punctually at 9:00 a.m. like it did every Saturday morning. Martin's father believed in keeping to a schedule, even when it came to social calls to his son. Martin picked up the receiver: "Hello?"

"Hi son. How was your week?"

"Oh, hi Dad. Pretty good. Studies were OK. Work at Tripos was a little busy, though."

"So how are things in the metal polish business?"

"Decent. Good enough to buy a new computer system. I installed it a couple of weeks ago. They're really glad to have a computer science major like me. I've been doing this for a year now, and I like it. But since my course load will be light next semester, I'll have more time to work for them. So they're also getting me involved with customer development. I go out with our sales manager on local trips to help him out. Also, on Fridays, they have me helping with inventory. That's when a lot of our delivery trucks go out. We get a lot of raw materials delivered then, too."

"Friday afternoons, you say?" queried the father. "I was hoping you could get time off on some of those. I've got some weekend business trips during the next few months. They're in Venezuela. I have to leave Friday mornings, usually. You know how you're mother isn't well these days . . . she's still carrying around the oxygen tank from when her emphysema got so bad. I was hoping you could drive down here from Penseroso and look after your mother when I'm gone."

"I guess . . . I get off work at 5. It's about a two-hour trip, with the rush-hour traffic here in Exodus. That brings me home at 7 on Friday at the earliest. Then I could spend the rest of the weekend."

"But then your mother would be alone for most of the day on Friday. Can't you shift your work hours? Say you got out at 3 or so to avoid the traffic. Then you could be home by 4:30 or 4:45, in time to help make supper. You know your mother can't cook when she's hooked to that tank. It's too dangerous."

"But Dad, it's not that simple!"

"I don't see why not. You told me last year that Tripos had a pretty liberal flextime policy. Taking care of your sick mother seems like a good reason to request it."

"Dad, that policy is for permanent workers. Most of the employees here have been around for a long time and are hard to replace. I'm temporary. There's no policy for people like me. Plus, a lot of these deliveries I'm supposed to track come in late in the afternoon. It's not like I can do my work just any time. I have to be there right then."

"Inventory isn't that hard to do," Martin's father persisted. "Can't they get someone to fill in for you? Or maybe you could just finish the job the next week. I mean, we're talking about your sick mother, here. As her son you have some responsibility . . ."

"Well it's *your* schedule that's the problem!" Martin broke in. "Maybe *you* can rearrange your schedule."

The elder Diesirae's voice hardened. "I can't do that. My flight schedule won't allow it. I need to have dinner on Friday nights with some of my clients. I'll just have to tell your mother you're too busy to take care of her."

"That's not fair!" Martin cried. "This job is important to me, and for more than just the money. I care about Mom as much as anyone! But it's not so simple to do what you want!"

◆ What should Martin do?

CASE 2.3 Conflict of Commitment

Myra Weltschmerz sighed wearily as she roused herself from a mid-afternoon nap to answer the telephone. She had a pounding headache and felt sick to her stomach, as she did from time to time. Normally she would not answer the phone in this condition, but she expected a call from her boyfriend Martin about supper plans that night. "Hello?" she mumbled.

"Hi Myra, it's Dolores."

"Uh . . . oh. Hi, Dolores." Myra could not conceal her disappointment.

"Is something wrong? You sound awful."

"No, just a little under the weather. I'll be better in a day or so."

"Oh, good. I won't keep you long if you're feeling bad. I just wanted to ask you to baby-sit my kids this Saturday. I have a job interview in Littleville at 4:30 p.m. It's for a small family-owned company there. I can't get off my present job during the week, but the owner was nice enough to see me on Saturday."

"That's good! What kind of work is it?"

"Oh, just some bookkeeping and stuff. But the best thing is that it pays a lot more than what I get now. And has more flexible hours on top! There's a small college nearby—maybe I can take classes there and finish my college degree. You know, the one I never finished when I married my ex-husband."

"Littleville . . . I forget where that is. Pretty far, isn't it.?"

"Yeah, about an hour and fifteen minutes. So I want to leave by 3 at the absolute latest. I'll be back around a quarter to seven if the interview takes an hour."

Myra's stomach sank. "I'm sorry Dolores, but I thought I told you—Martin and I are supposed to go to a play that night. It starts at 7, and he wants to take me to a romantic dinner beforehand. We haven't done too much of that this semester with our busy schedules."

Dolores's voice instantly betrayed desperation. "Oh, it really is this Saturday? I forgot. I thought maybe you said next Saturday! Are you sure about the time? I really need someone!"

Myra suddenly felt very nervous. She disliked it when Dolores got this way. "Yes, I'm sure. The play starts at 7. Can't you get someone else?"

"Well, I usually call you first. But because I couldn't remember exactly when you said the play was, this time I called the three other sitters on my list. They're all booked solid—no chance. Myra, can't you help me?"

Myra's headache intensified rapidly. "How about your ex?"

"He says it's not his weekend—he had them last weekend. I'm pretty sure he has no real conflicts, but he always gives me trouble when he knows I'm in a bind like this. That's one of the reasons I divorced him—he always knew how to take advantage." Dolores's speech quickened. "Isn't there, like, some way you can go out after the play? I get back by quarter to seven, so you should be able to make your play. Myra, this job might be the way to dig out of the hole I'm in—financially and otherwise."

Myra's voice quavered. "Dolores, you're putting on too much pressure. You make it sound like your whole life depends on me! I'm just a baby-sitter! Martin and I haven't done a romantic dinner in a long time. I promised him to wear my best dress, and I've really been looking forward to the whole evening. The play doesn't end till 9, and that's too late for a nice dinner. And even if we did the switch, fifteen minutes between the time you walk in the door and the start of the play is too short. How can I be sure you'll be on time? And we have to drive to the theater, park the car . . . you know! This is a fancy place—they don't let people in if they're late."

Dolores began to beg. "But Myra, remember last summer after I got laid off from that factory? You felt sorry for me and said that any time I really needed help, I should call you. Now I need help, and you tell me you want to go on a dinner date!"

These words cut Myra like a razor. She herself had been let down more times in her life than she could count. She had indeed made the promise. Myra began to shake, and nearly dropped the receiver. "Can't you change the time of the interview?"

"Myra, the owner is being incredibly nice as it is. He needs to fill the position right away, and I think he's talking to other candidates. He told me his schedule that day—I don't remember it all, but it sounded pretty full. I'm sure there are a lot of other people he can hire if I don't do what he asks."

◆ What should Myra do?

CASE 2.4 Reporting Unethical Behavior of Coworkers

"Got a question?" Terence Nonliquet inquired as he gathered his notes. He wished he could remember the name of the strikingly attractive student approaching him, but he had not yet mastered all the names of the students he was teaching. The Comp Sci 109 quiz section had just ended, and the room had emptied.

"I know I'll get a bad score on today's quiz again," Celia Peccavi responded. "That will be the second one I've messed up. I just can't figure out what the professor wants. Does he tell you TA's?"

Terence grew concerned. "No, he doesn't. But today's quiz was right out of his lecture Wednesday. Don't you remember?"

She paused and reflected. "No, not really," she confessed.

"He spent about ten minutes on it. I saw his notes. Although sometimes he gets hard to understand," Terence responded sympathetically.

His tone livened up her demeanor. "Yeah, I always understand you better than him." She paused and eyed him. "I think everyone in this section likes you. You say things clearly, and you really seem to care that we learn."

Terence looked down sheepishly. "Just doing my best." Looking up, he continued, "So is this stuff too hard for you?"

She shook her head and then tossed her hair over her shoulder in front of her. "No. I'm just not concentrating enough."

"Is something distracting you?" His concern was clear and genuine.

"Uh-huh." She paused, twirling her hair about her fingers. She glanced away out the window. "But I shouldn't bother you with my headaches."

"I can't help if you won't tell me," he responded with ardent concern.

A fleeting smile crossed her face as her glance returned to meet his. "Well, you won't tell anyone about this, right?" He nodded. She continued, "You see, it's like this. I work nights at Pandarus Pizza. You know the place? One of my coworkers has been taking food home with him. His name is Todd. We don't get along too well, because he's only concerned about himself. Plus he sort of hassles me sometimes."

"How do you know he's taking food?"

"I see him stuff it into his backpack. He's done it several times, although he doesn't know I saw him."

"Have you confronted him or told the boss?"

"No. I'm not sure what to do. A couple of months ago Todd accused me of giving free food to my friends. It wasn't true, actually—they had paid for it. But I can see how it looked that way. Anyway, he didn't tell anyone at the time. But now if I say anything about his taking food, he might accuse me in return."

"Why didn't he tell anyone if he thought you were giving things away?"

"I don't know. It doesn't matter, though. He's stealing, and I feel like I can't say anything. I think about it a lot, and now it's affecting my coursework." She paused, gazing into Terence's eyes. "What do you think I should do?" she implored.

◆ What should Celia do?

3

ANALYZING EXTERIOR ACTS: SOME FIRST STEPS

"General propositions do not decide concrete cases."
OLIVER WENDELL HOLMES (1841–1935)

Ethics as a Craft

The virtues are habitual ways of acting that are morally good. In simple moral situations, the virtues make right moral action almost effortless. More complex cases present more difficult challenges, however, and require much more effort and skill to fashion a satisfactory solution. The endeavor resembles that of a master craftsperson who constructs a piece of elegant furniture. The furniture maker uses many tools: hammer, saw, drill, and so on. In ethics, moral principles are the tools. However, tools by themselves do nothing. Making a table requires not only tools but also an understanding of how to use them. That is, the furniture maker needs a set of methods. The set should include not only the specifics for each tool, but also a general approach that brings these specifics together. In the same way, moral principles by themselves do not solve ethical problems; principles must be applied in the right way to particular situations through methods, both specific and general.

The analogy holds in a broader sense as well. The entire ethical life represents a craft like furniture making. Each moral action corresponds to a piece of furniture. Some actions require little more knowledge or effort to put together than a doorstop. Any apprentice could handle the job. Indeed, master furniture makers assign their apprentices many such tasks so that skilled use of simple tools and methods becomes habitual. In the same way, life hands each of us many simple moral tasks when we are young as practice to improve the virtues. More difficult tasks require more specialized tools, however, along with more sophisticated methods.

This chapter begins to develop more specialized methods. The general scheme represents an approach that most of us probably use already, at least intuitively. In using this scheme, however, we should remain cautious about how close to objective right we can come. It's not clear that every situation has a single "best" solution. In furniture making, there are several good ways to make a table. Whether "best" solutions always exist in ethics is a question for philosophy or religion. Nevertheless, some ways are better than others and some ways do not work at all.

Distinguishing Exterior and Interior Morality

Moral action has both exterior and interior dimensions. The exterior dimension concerns the goodness of actions viewed from outside the person. The interior dimension concerns the internal attitudes toward the virtues and includes personal intention.

To make this distinction clear, consider the following example. Suppose your sister has just been dumped by her boyfriend, and she calls you late one stormy evening looking for a shoulder to cry on. After hearing her story, you decide that your shoulder is better offered in person than over the phone. So you volunteer to drive over to her apartment in the small town 50 miles away. Wishing to reach her faster, you take a small county road that cuts twenty minutes off the route via the main highway. Darkness shrouds the road, however, and a small bridge ahead of you has washed away in the heavy rain. Despite your familiarity with the road and your best efforts to drive carefully, you fail to see the missing bridge and plunge into the creek below. The crash demolishes your car and lands you in the hospital paralyzed from the waist down. Did you do the "right" thing in deciding to take the small county road?

From the perspective of interior morality, your intention to rush to the aid of your sister seems good. Moreover, you exercised prudence by choosing a familiar road and driving carefully. However, in an external sense, the choice was "wrong." A broken bridge blocked your way, and driving off broken bridges generally causes trouble. It makes no difference that you didn't know about the broken bridge. You remain just as paralyzed as you would if you had driven recklessly off the bridge in a drunken stupor, intending to rub salt in your sister's wounds rather than tend to them. From your spinal cord's point of view, your intention, prudence, and state of knowledge make no difference.

In this case, a good moral decision from an internal perspective led to a bad external result. The reverse can also happen. How many of us have suffered under an egotistical teacher, coach, or parent who ruthlessly demanded our best performance in study or sports? In the face of this pressure, we decided to become first-rate students or athletes with bright prospects for the future—a good final result on balance.

Clearly there is a split that can divide exterior and interior morality. In this chapter, we will focus on analyzing the exterior part.

Beginning Case Analysis

We can begin moral analysis of a case by suggesting answers to the standard questions a journalist might ask: "Who?" "What?" "When" "Where?" "Why?" and "How?" Let's see how to begin to handle these by examining a particular case.

CASE 3.0 Being Told to Do Shoddy Work

Pandarus Pizza had a reputation for making the best-quality pizza in the city of Exodus. The restaurant was within walking distance of both Penseroso and Nosce te Ipsum universities, and enjoyed a loyal core of customers from both. Unfortunately, the onslaught of customers after football games of the Penseroso Peacocks sorely tried that reputation.

Celia Peccavi was learning that firsthand. Her eyes smarted from the smoke that spewed from the oven in front of her. It was a busy football evening at the restaurant, and in her rush to thrust a pizza into the oven she had spilled some of the topping, then dropped the pizza itself. What fell into the oven billowed smoke. She cursed under her breath. She hated this job, but needed the money for tuition.

The manager, Thorne Mauvais, rushed up behind Celia. "What's going on here?" he demanded. "We've got people who've been waiting an hour for their food!"

"There are too many orders!" Celia whined as she and some coworkers stooped to clean the mess. "The ovens only cook so many pizzas at once. If you go too fast, you start making mistakes."

"We've got to serve these football customers—that's where this place really makes money," Thorne scolded angrily. "I'm going to see that they get service."

"There's no way to go faster! We're doing the best we can!"

"I don't believe it. I've been here a long time, long before any of you. I've seen better throughput! If you're too stupid to quit spilling things, then take a few minutes off the baking time."

Celia looked up in shock. "Thorne, you can't do that. The pizzas need to be baked completely. They're not as good if they're not done through!"

"Never mind. These people are so hungry they'll eat anything. A lot are from out of town anyway. They have no way to compare what's better or worse. I'm telling you, take three minutes off, at least."

"Thorne, the middle won't get warm enough. You have to kill germs, if nothing else. We can't give people tainted food!"

"I told you, three minutes, at least!" Thorne growled.

Celia drew herself up and crossed her arms. "No way. I don't work here to put out junk. This place has a reputation for the best pizza in town, because we take time to do preparation right. That's what people come here for. I can't believe the owner would stand for this."

Thorne looked around, uneasily eyeing the other workers who were staring at him. "What are you all standing around for?" he demanded. "Get back to work!" They quickly complied.

Thorne drew up close to Celia, backing her nearly into the oven. He got so close she could feel his breath on her face. He looked around and whispered harshly, so only she could hear: "Look, I'm in charge here! You do what I say, or you don't work here at all! I do the hiring and firing!"

"Yeah, just like you take bribes, too," Celia responded with contempt. "I still remember . . . those football tickets you took from that window painter who wanted our business so bad. Maybe I should tell the owner about that too!"

Thorne could barely restrain himself. "One more word and you're fired!" he rasped. "And if you make a peep to the owner about any of this stuff, I'll make your life such a living hell that you'll beg to leave!"

◆ Should Celia follow Thorne's instructions about cooking? Why or why not?

◆ Should Celia tell the owner about the incident? Why or why not?

In analyzing this case, let's start with the question "who." A simple way to answer involves constructing a small table that assigns a row to each person or group of people involved. Here, obvious entries include Thorne and Celia. The case text also mentions Celia's fellow employees, the owner, and the customers in the restaurant. Finally, Celia refers to customers in general.

At this point you might ask, "Well, how do we decide who really needs to be listed? We could list virtually everyone on the planet!" The table can help by listing in its next column the interests each person has in the outcome. For example, all the characters have an interest in ensuring rapid preparation of safe, high-quality pizza. The restaurant owner and employees have interests in maintaining safe, friendly working conditions and keeping a public reputation for quality. Celia has an additional interest in maintaining her job to pay college bills, while Thorne has an interest in maintaining his authority.

The question "who" sometimes has another aspect, however. Sometimes one or more of those involved have characteristics with a moral dimension. For example, we expect especially high standards of conduct

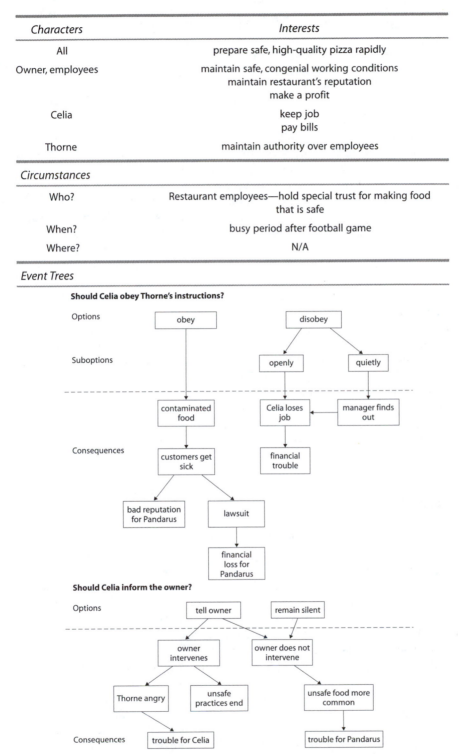

Characters	Interests
All	prepare safe, high-quality pizza rapidly
Owner, employees	maintain safe, congenial working conditions
	maintain restaurant's reputation
	make a profit
Celia	keep job
	pay bills
Thorne	maintain authority over employees

Circumstances	
Who?	Restaurant employees—hold special trust for making food that is safe
When?	busy period after football game
Where?	N/A

Event Trees

Should Celia obey Thorne's instructions?

Options: obey / disobey

Suboptions: openly / quietly

Consequences:
- contaminated food → customers get sick → bad reputation for Pandarus / lawsuit → financial loss for Pandarus
- Celia loses job ← manager finds out → financial trouble

Should Celia inform the owner?

Options: tell owner / remain silent

- owner intervenes → Thorne angry → trouble for Celia / unsafe practices end
- owner does not intervene → unsafe food more common → trouble for Pandarus

FIGURE 3.1 *Character table, circumstances, and event trees for Case 3.0*

from public officials, clergy, and physicians because of the special trust they hold. Characteristics of this kind form important components of what moralists call "circumstances." Circumstances occupy a central place in moral analysis because they can affect how good or bad an action is, even to the point of changing a good action to a bad one (or vice versa). Thus, it would be worse for a church pastor to shoplift a compact disk than for a teenager because most people expect clergy to set a higher standard of behavior. Circumstances can also affect moral responsibility for an action. For example, if the person stealing the compact disk were a toddler instead of a teenager or church pastor, the degree of responsibility would go down because a toddler would not know any better. In our analysis of cases, we'll list circumstances separately from the table of interests as shown in Figure 3.1. In the present case, all the restaurant employees, and especially Thorne as manager, hold a position of special trust to deliver food that is safe to eat.

Next, let's ask the questions "when" and "where." Time and place can clearly make a big difference in shaping the morality of an act, and therefore also qualify as circumstances. For example, shouting in a football stadium is usually appropriate, whereas shouting in a library is not. Reading the newspaper before bedtime is usually appropriate, whereas reading the paper during a final examination is not. In asking these questions, we should try to avoid a long listing of irrelevant details. Instead, we should list only those aspects of time and place that give moral coloring. In the case at hand, the hour and day are important because Pandarus Pizza is especially crowded after football games, which causes the stresses that fuel the conflict.

Event Trees

To continue our analysis, it helps to list the courses of action available to whichever character needs to make a decision. In this case, Celia has several options available to her. We can list these by asking, "*What* can Celia do?" and "*How* can she do it?" A little thinking quickly reveals that we should really look at these questions together because people can choose to do things on many different levels. For example, on a simple level, we can choose whether to use our turn signal when changing lanes on a highway. The act takes place in just a few seconds, with little opportunity for variation. "What" entails using the turn signal or not, while "how" becomes irrelevant. On a more complex level, we can choose whether to major in a field of study that will train us to help with the family business. Here, "what" might concern majoring in business versus music. If we choose music (not helpful for the family business), "how" might include suboptions like pursuing a minor in business or taking an extra course in accounting.

Put another way, some moral options specify simple actions. In such cases, we can construct a diagram having entries at its top for the main options that answer the question "what." For more complicated actions with several possible suboptions, we can draw in entries under the major headings listing suboptions that answer the question "how." We can link each suboption to its heading by an arrow as shown in Figure 3.1., which outlines what Celia can do. Notice that this situation actually requires Celia to decide two things. First, she must decide how to cook the pizzas. The main options involve obeying or disobeying Thorne's order to reduce the baking time. Second, she must decide whether to inform the owner. While the second decision really has only two possibilities, the first one offers several suboptions. In particular, if Celia disobeys, she can do so openly or secretly.

Now let's examine another dimension of the question "what." More specifically, " *What* consequences might follow from each possible action?" The approach we will adopt here has been employed rigorously in decision theory as applied to a variety of disciplines—artificial intelligence being just one. However, the approach does not appear to have made its way formally into the study of classical ethics. Hence, what follows represents an addition to classical virtue theory. In this approach, each suboption leads to events that we can draw as shown in Figure 3.1. Inclusion of the consequences turns Figure 3.1 into an "event tree,"[1] which shows clearly in pictorial form the cause-and-effect relations among options, suboptions, and consequences. Event trees (and their more quantitative cousins, decision trees) appear to have originated in the work of the famous physicist Christiaan Huygens in the seventeenth century,[2] and are employed today by statisticians, econometricians, decision theorists, and social scientists.

Notice how the tree links suboptions leading to identical consequences. Such linkage is not required, but this kind of shorthand helps prevent writer's cramp! Notice also that a given event can lead to two contradictory consequences. For example, if Celia tells the owner about Thorne's order, it's not certain whether the owner will intervene.

In constructing event trees, we should avoid entering consequences that would have occurred even in the absence of the action. For example, if Celia obeys Thorne's order, it is virtually certain that she will keep her job. However, she would have kept her job anyway if the incident had never occurred. As obvious as this idea may sound, the temptation to record opposite consequences for opposite actions can be very strong. For example, since disobeying openly leads to firing, it seems natural to record that obeying leads to job retention. However, firing would not have happened without the incident, whereas job retention would have. Avoiding entries like job retention keeps the event trees more manageable, helping to avoid confusion. Deciding at what point to stop listing consequences in the tree remains largely a matter of judgment. However, very often the tree is limited by our ability to predict the future.

Commonly the event tree has options with both good and bad consequences. These options present moral conflicts for exterior morality. However, for the moment we have left aside the journalist's question "why." This question connects with intention and interior morality, which we will examine in the next chapter.

A REAL-LIFE CASE: Dow Corning Corp. and Breast Implants

Dow Corning Corp. first began marketing silicone-based breast implants in 1964. At that time, no evidence existed to show that silicone materials caused health problems. Indeed, various short-term studies performed by one of Dow Corning's parent companies, Dow Chemical, indicated that silicones were biologically inert. However, no law required government approval or extensive testing. By the mid-1970s, Dow Corning had evidence that the implants could leak or break, and therefore began a new set of tests for health hazards. Most of the silicones continued to show biological inertness. However, one type of low-molecular-weight silicone, present in very small quantities in implants, appeared to stimulate the immune systems of mice. Dow Corning did not publish the study, and continued to market the implants, even in the absence of long-term data on health safety.

Throughout the remainder of the 1970s and into the 1980s, reports of leaks and breaks mounted, together with various reports of fevers, headaches, and immune system problems. The Food and Drug Administration (FDA) gained the ability to regulate medical devices in 1976. Implants were allowed to remain on the market, since no coherent pattern of problems had become apparent. Up to two million women received them over a period of nearly three decades. Serious problems did come to FDA's attention in the early 1990s, however, when lawsuits against Dow Corning by women with implant problems began to mount. In 1992, FDA decided that Dow Corning had not done enough to demonstrate implant safety, and forced the company to stop selling the devices except for clinical studies and for women with clear health needs. The issue became an instant sensation in the news media, and in the face of a flood of litigation and civil judgments, Dow Corning declared bankruptcy in 1994, effectively freezing the lawsuits. Litigation continues at the writing of this book, although there is hope that the major class-action suit will be settled shortly.

Vigorous disagreement has continued on several fronts. Several aspects of the case complicate ethical analysis. "Silicone" is not one chemical but a whole class of them. Implants contain a lot of some silicones, tiny amounts of others, and none of still others. The activity of these various compounds in humans (as opposed to in laboratory animals) is unclear. Epidemiological studies consistently find no connection between implant use and human disease, but these studies have significant limitations in detecting low-incidence problems and in establishing cause and effect. Women reporting implant problems often tend to exhibit nonspecific symptoms like fever or headache whose causes remain unclear. Even the experts at FDA and the American Medical Association have been at odds over what the availability of silicone implants should be. The "facts" of the case tend to vary depending on who is asked—not surprising because large amounts of settlement money (billions) are at stake. Scientific truth and ethical fairness are difficult to discern amid the legalese of the courtroom and in the sensationalism of the press, fanned by partisans on all sides.

- To what extent should a medical product be proven safe before use? Does it matter what the end use is (a potentially life-saving AIDS drug, for example)?
- What should be done about products that were first marketed in a different regulatory environment?
- What should Dow Corning have done at the various stages of this episode?

References

Byrne, John. *Informed Consent.* New York: McGraw-Hill, 1996, 15–40.
Vassey, Frank. *The Silicone Breast Implant Controversy: What Women Need to Know.* Freedom, Calif.: The Crossing Press, 1993.

"He who wears his morality as anything but his
best garment were better naked."

KHALIL GIBRAN (1883–1931), *THE PROPHET*

Notes

1. For a detailed discussion of event trees, see Glenn Shafer, *The Art of Causal Conjecture* (Cambridge, Mass.: MIT Press, 1996), ch. 2.

2. Ibid.

Problems

1. Write a paragraph or two describing an ethical dilemma you have encountered in a job you've had. (If you've been lucky enough never

to have been confronted with a problem like this, describe one that a friend or relative of yours has had.) Analyze the case as follows:

   a. Begin by writing down in tabular form the people involved and the interests each has in the case.

   b. List any circumstances that have significant relevance to the moral decision at hand.

   c. List the options/suboptions available to the person who has to make a decision, together with the event tree flowing from each option.

   d. Recommend what you think the person should do. Note: you don't have to say what was actually done in real life (unless you want to)!

2. Each case below has a question after it.

   a. Begin by writing down in tabular form the characters and the interests each has in the case.

   b. List any circumstances that have significant relevance to the moral decision at hand.

   c. List the options/suboptions available to the main character who has to make a decision, together with the event tree flowing from each option.

   d. Recommend what you think the character should do.

CASE 3.1 Going over a Superior's Head

Terence Nonliquet sniffed the crisp air of early autumn and adjusted the strap on his helmet before mounting his bicycle. His girlfriend Leah Nonlibet waited somewhat impatiently on her bicycle a few feet in front of him. He filled his chest again with the fresh air and said, "It looks like a great day for a ride. Maybe we can do 30 miles instead of 25!"

Leah scowled and replied, "I doubt we'll have time. Your mother's phone call took a long time, and I have to get back in time to finish a lab report this evening. And you know how I hate to get to bed too late on Sunday night."

"Oh," he responded. Silence reigned between them for the first three miles. They stopped at a red light even though no car was coming. Terence remembered the argument he had had on that subject with Leah the week before. She stopped for red lights come hell or high water. Terence's easy demeanor slowly became more tense as he replayed the scene. Then other unpleasant thoughts began to absorb him.

Noticing how quiet he was, Leah inquired, "Is something wrong? I mean, there's no problem with doing 25 miles."

"It's not that. It's the computer science class I teach. I've been doing this job now for about six weeks, and I'm learning that I don't like the head instructor too much."

"You mean Professor Bligh?"

"Yeah, I don't think he's very fair. Like this exam we gave on Friday. It was the first hour exam. I heard a rumor from one of the students that Bligh doesn't change his exams too much when he teaches this course. I asked him the day before class about this, and he just shrugged it off. 'It's hard to give very original problems in a lower-level course like this,' he said. 'I change some of the details each time. And I only teach it every two or three years, so there's not too much talking between the students of different classes.' Leah, I'm only a junior myself, and I've never written an exam before, so I don't know how hard it is. But just because different classes don't talk doesn't mean there aren't files out there in fraternities—you know."

"I can understand what he means, if he changes details each time," Leah offered sympathetically.

"But that's the real problem—I don't think he does! And since this exam I'm sure of it."

"How can you be so sure?"

"The average was about 70 percent, and the distribution in my section looks pretty normal except for three students. They don't come to class too much, and their homeworks are below average. Still, they got scores above 95 percent!"

"Maybe they copied each other. Were they sitting together near a smart person?"

"No, they all sat apart. And get this—as I was walking back from class after the exam, I came up behind one of them who was waiting for the light to change at a street corner. He was talking to a friend who was not in the class, and bragging about what a piece of cake the exam was because it looked *exactly* like an old exam in his fraternity file."

"So Bligh was lying!"

"I think so. And it's not fair to the students who don't have the files."

"What are you going to do?"

"I don't know. Bligh basically blew me off once already. If I bring it up again I'll pretty much have to call him a liar."

"But he is!"

"Yeah, but he's also my boss. He recommends whether I get this job again next semester. He can also assign a lot of the course grunt work to me, like writing up the homework solutions. That takes a lot of time, and I'm overloaded this semester already. Right now he assigns things like that pretty arbitrarily—it's not always a fair distribution of labor among the TA's."

"Why don't you tell the department head?"

"Professor Peccavi? He's always so busy. I hear it's really hard even to get an appointment. He's got other things to think about, and probably won't like some undergrad TA coming in to complain. If I get a reputation as a troublemaker, he won't want to approve me as a TA next semester, even if he ignores Bligh's recommendation letter. Plus, I don't really have any proof—only hearsay. I think faculty look out for their own more than the students anyway."

"But you don't know that. Have you ever met him?"

"Only once. He seemed nice, just busy. I don't want to take unnecessary chances, since this job looks great on my resume, and I can use the extra money."

◆ What should Terence do?

CASE 3.2 Doing Private Work for the Boss

Leah Nonlibet stared blankly into the dishpan of the lab sink where she was cleaning glassware. The rest of her job as an undergraduate laboratory assistant excited her greatly, but this kind of drudge work sapped her energy, especially on a Friday of a week as hard as this one had been. Professor Warren Clark walked up behind her. "It seems like you've been washing more dishes than usual lately, Leah," he said.

"Yeah, and it's pretty boring."

"I'm sorry about that. I've always wanted to get a dishwashing machine for my lab, but there's not a lot of money for that kind of thing in the field of geology. It doesn't help to be at NTI, either. We're really not a major research university—most of my colleagues don't do much research at all. It's a lot harder to get grants funded here. So I have to resort to hiring undergrads. Anyway, I think the dish load will drop in about a week as Bernard finishes his experiments on the mass spectrometer. It's not working too well, you know, so I'm letting him start another project for the rest of his MS thesis."

"It's OK—I'm glad Bernard will get something new to do," she said.

"Listen, Leah, I have a favor to ask. I want to take my wife to a special surprise birthday party tomorrow night. When we go out, I usually get the woman next door to look after my daughters. But she's busy tomorrow, and I haven't been able to find anyone who can fill in. I know you're very responsible and that you like kids. Would you be willing to watch them? I'll pay you whatever you think is fair."

"Professor Clark, I thought your daughters were old enough not to need a baby-sitter any more."

Professor Clark turned sheepish. "Well, they're 16, 15, and 13. They should be old enough. But they get pretty unruly sometimes, and quite honestly I really can't trust them alone. The oldest can drive now. I've caught her taking the car out without permission when my wife and I are gone. The middle one likes to be with her new boyfriend a lot, but I don't like him and don't want him in the house when I'm gone. He likes to drink beer, and I don't want her starting in with that. The youngest likes to cook—but she's careless. Last week she started a grease fire on the stove because she didn't watch what she was doing."

"Well, I don't know," Leah intoned slowly. "I promised my boyfriend Terence to see a movie with him tomorrow night."

"You're free to bring him over. We have a top-notch VCR and big-screen TV, with a big library of movies. But if there's something special you want, go ahead and rent it, and I'll pay for it."

"I'll talk to Terence, and get back to you by suppertime, OK?"

"Sure. But I really hope you can do it—I'm in a real bind."

Fifteen minutes later, Leah had reached Terence on the phone and had related her conversation with Professor Clark. Terence was not happy. "We agreed we'd see Star Wars VIII together. A real movie . . . in a theater. Not a video. Spending the evening with three immature teenagers isn't my idea of a hot date!" he grumbled. "And you're the one who's been complaining we don't spend enough quality time together! If he has such wild daughters, that's his problem, not yours."

"But Terence, we didn't pick any particular movie. And he's my boss! He's been really good to me, too! A couple of weeks ago when I complained to him about how Marcus's girlfriend was always calling and tying up the lab phone, Professor Clark was very sympathetic to me and talked to Marcus about it. Since then the problem has improved a little. Anyway, I want a good recommendation letter from Clark next year when I start looking for graduate schools. It's only this one time he wants me to baby-sit."

"How do you know that?" Terence retorted. "If you do a good job, he'll want you back next time. If you can't say 'no' now, how will you do it then?"

◆ What should Leah do?

CASE 3.3 Gender Issues

Emily Laborvincet's heart pounded with anticipation as she made her way across the work floor of the Tripos Metal Polish factory. She had some misgivings about rejecting her offer at Ajax Foods after first accepting it, but the rich promise of spring-semester work-study at

Tripos soon suppressed those thoughts. Tripos had also offered her some part-time work for the autumn, and today was her second day. She had spent her first day meeting coworkers and learning about the operation. Tripos employed only about fifty people, most of them hourly workers on the factory floor. The company owner, Monica Ichdien, liked to draw some of her white-collar help from Penseroso University, since the students there proved bright, flexible, hard-working, and inexpensive. Monica also liked working with idealistic young students. Emily's job would begin with some bookkeeping and accounting, and would eventually entail some production schedul-ing—heady stuff for a sophomore in chemistry!

Just now she was heading for the accounting office to check out the computers and software the company maintained. While looking up at a big mixing tank as she passed, she accidentally ran smack into Floyd, the hulking, graying floor manager for half the factory. Emily bounced off Floyd and stammered, "I'm sorry . . . very sorry, Lloyd," as she recovered her balance.

"It's Floyd, not Lloyd. And ya need to watch where yer goin', young lady," he growled testily. Then his eyes narrowed. "Yer the new girl, aren't ya?"

Emily stiffened. "My name is Emily. It's my second day. We met yesterday, I think."

Floyd drew himself up to his full 6-foot 4-inch height. "Yeah, I re-member. Yer the girl Monica said would do the accounting. She said ya'd manage scheduling, too."

"Yes."

"Well, if ya do scheduling, you'll deal with me. I'll tell ya some-thing. I never liked it when Monica started hiring college students for that. They know all about books, but nothin' real. And the last girl we had—gawd! A good looker, mind ya, but didn't know a 3 from an 8. I been here over thirty years! I showed *her* who was who. She learned to stay outta my way."

Emily's stomach tightened, and her eyes fell to the floor. "I'm sure we'll get along," she murmured.

Floyd sensed he had the upper hand. "Yeah. Just remember, I don't take no orders from no girl. Monica, she's OK. She's almost my age, and knows her way around. But no girls!" With that he turned and stalked off.

That night, Emily went over to her boyfriend Todd's dorm room to study, as was her custom. But she could muster no concentration, and fiddled with her pen for about an hour. Finally Todd looked up and said, "Are you going to study tonight or what? You're just play-ing!"

"I just can't concentrate. Something happened at work that got me upset."

"Oh?" responded Todd. Then Emily related her encounter with Floyd. As she finished, she said, "I'm not sure what to do. The guy is really biased. He could make things tough for me. I already hate being around him."

"Maybe he's all bark and no bite," Todd remarked. "Sometimes guys, especially old guys, just have to say their piece. It's like a dog marking his territory. But if you don't go out of your way to stick your finger in his eye, he's OK."

"I don't think so. Not Floyd. Something tells me he means it." she answered uneasily.

"Well, if you really think so, then you have to come out swinging!" Todd exclaimed. "You think Floyd is biased. He sounds more like a bully to me. I know about bullies. Believe it or not, they don't actually like to fight." Emily's eyes widened in disbelief. "Oh, they'll fight if they have to," Todd continued, "but they don't like to get hurt. Even if they win."

"So what should I do?" Emily snorted. "Punch him out? He'll clean the floor with me!"

"I don't mean a fist fight. Here's what I mean. You're new at Tripos, new and inexperienced. That makes you weaker than Floyd. He'll try to push you around—call you a stupid girl and all that. If you get flustered, that makes his day. If you ignore him, he'll keep pushing until you respond. By then you've got a real problem. No, you have to bloody his nose right away. Make life uncomfortable for him somehow. Even if he 'wins,' he'll learn that every time he bothers you he'll have pain. Like any bully, he'll leave you alone pretty quick."

"And how do I make life 'uncomfortable' for him?" Emily demanded.

"That depends. In your case, I'd go right up to the owner—Monica, right?—and tell her what happened. Ask for some concrete action."

"But I'm new. I don't want to cause trouble! How do I know she'll back me? I heard her say that Floyd is the backbone of the factory floor!" contended Emily. "And I'm not going to call the Feds or file a lawsuit over something like this. I have other things to worry about in life, like homework!"

"Yeah, but you know Monica likes college students. And she's a woman, so she's probably dealt with what you face somewhere along the way. She wouldn't hire you if she didn't want you to succeed. You're right, she may not do all that much. But Floyd can never be sure, and he knows you won't take his nonsense lying down. That's the important point."

Emily pursed her lips, deep in thought.

◆ What should Emily do?

CASE 3.4 Computer Games on Company Time

Todd Cuibono and Celia Peccavi sat listlessly at one of the customer tables in Pandarus Pizza. Autumn was well under way, but the oppressive heat of midwestern summer was staging a final brief counterattack. Worse yet, the restaurant's air conditioners were shut down for the season, and the ceiling fans provided scant respite from the thick night air. Through the open window, the chimes of a nearby bell tower tolled 10. But aside from Todd, Celia, and a few idle workers in back, no one else was in the restaurant to hear.

"Not a customer in an hour," sighed Celia. "And another hour to closing time."

Todd nodded. "Yep, it's pretty boring. And I doubt we'll get any more business tonight. It's just one of those slow times."

Celia suddenly got up and stretched. "Well, I'm sick of just sitting here with nothing to do."

"You can always mop the floor again," laughed Todd. "Or clean the bathroom toilets." Then he sneered, "Don't a lot of graduates from NTI get jobs like that?"

Celia scowled. "NTI graduates do just as well as you Penseroso geeks. Anyway, this place is clean already. We've done all the busywork stuff. Say Todd . . . you sometimes do work in the back office. On the computer. Do you know if Thorne has taken all the computer games off the hard disk like the owner said?"

"What difference does it make?" muttered Todd. "Mr. Laird said he would fire anyone playing games on that computer. Remember? He got really mad after he walked in and caught someone on the afternoon shift doing that. And he told Thorne to enforce the rule with a vengeance."

"Yeah, but Thorne is sick today, so you're in charge," retorted Celia. "And Mr. Laird hardly ever comes in so late at night. So we won't get in trouble. Are the games still there?"

Todd shifted uneasily in his chair. "What makes you so sure I would allow computer games as acting manager?" he challenged Celia.

"Because you like them as much as I do. Anyway, we both know why Mr. Laird said what he did. Those games can be addictive. I heard people on the afternoon shift played them even when there were customers around. Mr. Laird just wants to be sure the work gets done first. But now there's no work. What harm can there possibly be in passing the time when there is nothing—and I mean nothing—else to do?"

Todd straightened up. "I have my responsibilities," he declared.

Celia looked Todd dead in the eye with contempt. "Get a life! Since when are you such a stickler for rules? I'm sure you don't follow them all the time."

"What do you mean?" Todd asked uneasily.

"For a while I noticed there were some desserts and breadsticks missing. I didn't say anything at the time." Celia paused for effect. Todd squirmed under her piercing gaze. She continued sarcastically, "You wouldn't happen to know where they went, would you?"

"I don't know what you mean," Todd murmured weakly. "Anyway, as far as I know Thorne has not removed the games," he added quickly. "Actually, I think sometimes when he goes in his office and closes the door, he plays them himself. He rarely takes what Mr. Laird says seriously unless he has to."

"Well, what are we waiting for?" exclaimed Celia in triumph.

- ◆ What should Celia do?
- ◆ What should Todd do?

4

ANALYZING INTERIOR INTENTIONS: SOME FIRST STEPS

"The intention makes the crime."
ARISTOTLE (384–322 B.C.), *RHETORIC*, BOOK I

In this chapter we begin to examine a question of key importance in moral analysis: "why," as in "*Why* might this option be chosen?" In other words, we look at the role of personal intention. For many centuries, moralists have attached supreme importance to intention in determining the morality of an action. It is not difficult to see why. We become much angrier with someone who deliberately drops our hand calculator on the floor than with someone who does so accidentally. In fact, our legal system accounts for intention by distinguishing between voluntary and involuntary manslaughter, for example, and between tax evasion and inadvertent tax underpayment.

Describing Intention

Intention can be understood in several different ways. Many moralists define intention as the purpose for which an action is done. Intention is the answer to the question, "What is this action trying to achieve?"

Since sometimes we do things with several goals in mind, an action can have several intentions. In such cases, it can prove useful to approach intention from a different angle. The goal or purpose of an action can almost always be described as an event. With some thought, we can describe even rather vague goals this way. For example, if we want to build a bigger vocabulary, we can restate the goal as wanting to increase our vocabulary by roughly three hundred words. Reaching this state is an event that happens at a particular point in time. Phrasing our goals in terms of events offers the advantage of making us able to identify systematically our intentions by determining which consequences of an action we want to happen.[1] That is, we can answer the question, "To which consequences do we give approval?" Some moralists employ the word "consent" instead of "approval." We will employ "approval" here because "consent" suggests a rather passive meaning that runs counter to the activism of a thoroughly moral life. Regardless the exact choice of words, the main point should remain clear: we can identify intention by examining attitudes toward the consequences of an action. We can usually classify such attitudes in one of four ways: approval, disapproval, mixed, and indifferent.

Let's examine more closely what these assignments mean, starting with "approval" and "disapproval." These attitudes differ from mere positive and negative feelings. We may feel bad when we tell an instructor about a good friend who has cheated on an exam, but we may nevertheless approve of the justice that follows. We may feel good when our friends eagerly invite us to a party, but we may disapprove of the drunkenness, drug abuse, and vandalism that we know this particular party will bring. Since power to choose lies in the will, approval and disapproval become chosen intentions of the will rather than involuntary responses of the emotions.

How about our attitudes that are neither positive nor negative? We can distinguish between "mixed" and "indifferent." The two attitudes are not the same. To remain indifferent implies moral insensitivity. Extreme cases include sociopaths who show no sorrow or concern after committing horrible crimes like murder. In the long run, a pattern of moral insensitivity leads to moral failure. Thus, we should try hard to avoid indifference. On the other hand, having mixed opinions may indicate a very refined moral sensitivity, which lies at the foundation of the upright life. Thus, while we should try to resolve mixed opinions where possible, they can be tolerated far more than indifference.

The Importance of Intention

But why should intention matter so much? After all, sometimes good people do bad things and vice versa. Sometimes well-intentioned social workers unwittingly drive themselves to complete burnout, hurting themselves and depriving needy citizens of needed attention. Inversely, sometimes brutal dictators build schools and hospitals in a cynical attempt to gain favor with the outside world. In the end, aren't the external actions all that really matter? Should we care about interior attitudes when exterior actions ultimately determine human well-being? While there is room for debate on this subject, it seems too simple to say that the overdriven social workers have done wrong and that the cruel dictators have done right.

An example might shed more light on this question. Suppose Greg has a classmate named Sarah, who in turn has a younger sister Clara. One day Greg happens to meet Clara briefly at a party. He falls in love with her, and asks her for a date. Wary of men who fall in love at first sight,

Clara declines. But Greg refuses to give up, and develops a clever scheme. He decides to pursue his classmate Sarah, who has shown romantic interest in him. Greg does not really like Sarah, but he calculates that by developing the appearance of a healthy relationship with her, he can maintain some loose contact with Clara. Clara will see that he's really not such a volatile person after all, and will come to trust him more. At that point, Greg plans to drop Sarah in a way that won't frighten Clara, freeing him to pursue the real object of his affection. So Greg begins to date Sarah, who falls in love with him. Greg still has no liking for Sarah, and carefully avoids any expression of romantic affection. He treats her respectfully, but limits dates to a weekly basis, and periodically hints that he does not want the relationship to progress too quickly. Sarah, however, interprets this reluctance as merely shyness. As Greg has planned, Clara observes the relationship and becomes impressed with his conduct. Greg sees this, and prepares to end the relationship by inventing some excuse about incompatibility with Sarah. However, before he can carry out his plan, Sarah's car is hit by a drunken driver, and she dies a tragic death. Clara despairs the loss of her sister, but Greg sees his opportunity. He provides a comforting shoulder to cry on, and after a period of mourning, Clara readily falls in love.

Should we approve of what Greg has done? After all, throughout the episode his external actions have shown few flaws. He has treated Sarah well and has not pressured Clara. It's difficult to charge Greg with outright lying, since like many men in love, he does not express affection openly. He allows Sarah to continue in her dreams of love, but provides carefully worded warnings against them. It's true that had Sarah lived until Greg dumped her, we could convict Greg of unfairly breaking her heart. But as events actually turn out, Greg gets what he wants without hurting either Sarah or Clara. Despite all this, most people would criticize Greg's approach. While his outward actions might look fine, and might actually imitate what a shy but upright man might do, his inward intention is very different.

This example highlights the exterior and interior aspects of moral action. Greg acts morally in an exterior sense, but his interior intentions are a different story. In fact, his service of external morality occurs by accident. Does this matter? Our approach to ethics based on virtues argues that it does. Virtue focuses on the theme of habit. That is, voluntarily doing something tends to internalize a behavior pattern that makes that action easier to do the next time. This process of internalization lies at the root of the old saying, "You become what you do."

Consider the analogy of a sickly 98-pound weakling who looks at himself in the mirror, becomes disgusted with what he sees, and resolves to improve his physical conditioning. He decides to maintain a balanced diet, run 5 miles per day, and lift weights. At first, his body responds with pain and exhaustion. However, after a year, his physical condition improves greatly. His muscles strengthen; his lung capacity enlarges; his endurance grows. He no longer dreads exercise, but instead looks forward to it daily. He sickens less often, and his mind functions more clearly. In short, he is no longer merely someone who does exercises. He has become an athlete—his behavior has altered the very nature of his being.

In the same way, doing something that has moral significance has the consequence of imprinting our behavior patterns. However, suppose that like Greg we intend an action that we are prevented from performing. Does the notion of imprinting still apply? After all, acting on a decision has more effect than merely thinking about it. To some extent, however, simply deciding also has importance. Recall that criminal law punishes conspiracy to murder as well as murder itself. Even if a plan for murder never sees completion, the planning itself imprints the conspirator in a way that makes killing more likely in general.

Of course, human behavior does not obey fixed mathematical rules. Under most circumstances, people can consciously control their behavior in spite of habit. Acting ethically resembles breathing. Normally we breathe without thinking, but within limits we can control our breathing if we choose to. Thus, in saying that one decision affects others that follow, we can speak only in terms of likelihoods, not certainties. This idea has been well established by the social sciences. A convicted criminal is more likely to commit a crime in the future than someone with no record. A one-time user of illicit drugs is more likely to use such drugs in the future than a zero-time user.

Getting back to the example of Greg and Sarah, we can criticize Greg on these grounds. By deciding to willfully manipulate and hurt Sarah, he becomes more likely to manipulate and hurt women in the future. While he never actually brought Sarah to grief, deciding to pursue this injustice makes him more likely to do it again.

Effort and the Virtues

We have been looking at intentions mainly in connection to specific decisions. We can, however, consider moral attitudes in a more general sense. There is a strong analogy between the moral life and athletics. A well-conditioned athlete climbs mountains better than a couch potato because the athlete habitually performs strenuous and complex physical tasks. However, the analogy also holds on a deeper level. Rarely can physical conditioning remain stationary at a fixed level. At any given time, an athlete's capabilities are usually either improving or declining. (Note that capability differs from what actually happens; a jogger can run at the fixed level of 5 miles per day for many years, but endurance beyond that distance may change enormously with time.) Sometimes the change occurs quite rapidly, while other times it may remain barely perceptible. If nothing else, things like illness, changes in lifestyle, and advancing age take their toll unless actively resisted. Change seems inevitable, and human nature has a nasty tendency toward backsliding in the absence of continual efforts to improve.

Experience teaches that similar rules govern the moral life. It makes little sense to say, "Well, I've reached a pretty high standard for moral conduct; now I can relax and let my ethical behavior take care of itself." It is true that acting according to the virtues in one case makes it easier to behave this way again. However, a certain amount of effort is always involved. Muscles weaken without such effort, and so does the moral will. Thus, if we just try to "get by" in the moral domain, without continually attempting to improve, we can usually expect to backslide sooner or later. Such attempts should focus not only on handling simple situations more consistently, but also on dealing with more complex ones.

These ideas have enough importance to merit a fundamental principle of morality:

Principle: *People should try insofar as possible to continue to progress in the moral life.*

That is, attempting to merely run in place should be avoided. Of course, this principle probably needs philosophical or religious arguments for its complete justification; we will not outline such arguments here.

The Role of Benevolence

The principle we just stated may sound rather bland, but it actually has potent implications for a morality that leads beyond the classical virtues. We have proposed that we should avoid doing just the bare minimum to get by in the moral life. However, does avoiding the bare minimum require us to practice the opposite, benevolence? Some argue that benevolence is the highest good; we should do what is best for others without any expectation of return. Others insist that the classical virtue of justice obligates us to do good only so far as we ourselves benefit. Clearly these two points of view can lead to very different conclusions. No one disputes that people whose chief concern is for others do great things for humankind and merit great respect. However, is this benevolence required of everyone, all the time? If benevolence is not required, are we really being just selfish?

Attaching benevolence to the basic principle we have been discussing has been debated for centuries within many philosophical and religious traditions. Many of those traditions promote benevolence as the highest, possibly supernatural, good. Most, however, allow that some amount of self-interest does not fatally poison a moral decision. In the language of our principle, the key words are "insofar as possible." This qualifying phrase translates into actual practice through the virtue of prudence, which sometimes recommends against heroic action given the basic weaknesses of the human condition.

A REAL-LIFE CASE: The Tuskegee Syphilis Experiment

In 1932, the U.S. Public Health Service (PHS) began a study of the progression of syphilis using a group of four hundred infected African-American men in rural Macon County of Alabama. The men were given transportation to hospitals, medicines (except for syphilis), and burial services if required. The study originated as a treatment program shortly before 1932, but because of lack of funds changed to an observation of the progression of the untreated disease. However, PHS did not tell the men that the study's purpose had changed. In fact, PHS never told the men they had syphilis at all, but instead referred to their illness as "bad blood." Although the men thought they were receiving treatment for "bad blood," in fact their medical care specifically avoided treating the underlying condition after 1932. During the early years of the study, there was in fact no cure for syphilis—only methods to treat the symptoms or slow the progression. In the early 1940s it was discovered that penicillin could cure the disease outright. Despite this discovery, PHS doctors did not change the study; the subjects were neither given penicillin nor told of its effectiveness for their disease. The study continued for forty years, until 1972, by which time over one hundred of the men had died from the disease or its complications. Many of the remaining men suffered from the heart disease, apoplexy, blindness, and insanity characteristic of advanced syphilis.

When these facts became known in 1972, detailed inquiries were made into what motivated the experiment. Several reasons came to light. First, early in the study there existed a clear need for establishing serologic tests for syphilis and for characterizing a rigorous control group as a benchmark for testing therapies in other groups of subjects. Second, after the effectiveness of penicillin was discovered the researchers wanted to preserve an untreated group for long-term study in a society in which most infected people would receive treatment shortly after diagnosis. Third, some PHS officials planned to compare the results of the Tuskegee study with those of a previous Norwegian study on whites and of projected studies on Native Americans and possibly other minority groups. These comparisons would presumably provide comparisons of the sexual practices of these various groups.

The Tuskegee experiment has left deep and lasting scars among African-Americans in many communities of the South. Considerable skepticism exists toward current programs for the treatment and prevention of AIDS, for example.

◆ To what extent were the intentions behind the Tuskegee experiment valid?

◆ What role do you think racial attitudes toward African-Americans played?

References

Benjamin, Roy. "The Tuskegee Syphilis Experiment: Biotechnology and the Administrative State." *Journal of the American Medical Association* 87 (1995):56–64.

Lee, B. L., R. Maisiak, M. Q. Wang, M. F. Britt, and N. Ebeling. "Participation in Health Education, Health Promotion, and Health Research by African Americans: Effects of the Tuskegee Syphilis Experiment." *Journal of Health Education* 28 (1997):196–200.

"He who moves not forward goes backward."

JOHANN WOLFGANG VON GOETHE (1749–1832), *HERMANN AND DOROTHEA*

Note

1. This approach meshes nicely with the idea of event trees developed in chapter 3.

Problems

1. Write a paragraph or two describing an ethical dilemma you have encountered in a job you've had. (If you've been lucky enough never to have been confronted with a problem like this, describe one that a friend or relative of yours has had.) Analyze the case as follows:

   a. Begin by writing down in tabular form the people involved and the interests each has in the case.

   b. List any circumstances that have significant relevance to the moral decision at hand.

   c. List the options/suboptions available to the person who has to make a decision, together with the event tree flowing from each option.

d. Add your assessments of intention to an event tree with a small notation by each consequence in the event tree. You can use shorthand notation like "a," "d," "m," and "i" for "approve," "disapprove," "mixed," and "indifferent." If there is no way to assess intention for some consequences, use a notation like "u" for "unknown."

e. Recommend what you think the person should do. Note: you don't have to say what was actually done in real life (unless you want to)!

2. Each case below has a question after it.

a. Begin by writing down in tabular form the characters and the interests each has in the case.

b. List any circumstances that have significant relevance to the moral decision at hand.

c. List the options/suboptions available to the main character who has to make a decision, together with the event tree flowing from each option.

d. Add your assessments of intention to an event tree with a small notation by each consequence in the event tree. You can use shorthand notation like "a," "d," "m," and "i" for "approve," "disapprove," "mixed," and "indifferent." If there is no way to assess intention for some consequences, use a notation like "u" for "unknown."

e. Recommend what you think the character should do.

CASE 4.1 Dating between TA's and Students

Celia Peccavi waited until her fellow students cleared the room after Friday's class before approaching her teaching assistant Terence Nonliquet, who was erasing the chalkboard. "Hi," she ventured timidly. "Do you remember last week when I told you about my coworker at Pandarus Pizza who was carting off food?"

Startled, Terence stopped erasing. "Yes," he replied. "How could I forget?"

Celia's voice strengthened. "Well, I decided to just keep quiet and not say anything to the boss. I didn't want to get into a lot of messy accusations, evidence, and all that."

"Oh," said Terence, returning to his erasing. "I guess that was probably hard to decide."

A pause followed. Celia glanced out the open window, then up at Terence from under the hair that spilled into her face. "I was wondering. . . ." Her voice faded slightly, then revived. "Well, my dorm floor is having this pig roast tomorrow over at Joie de Vivre Park. We'll have tons of food, and games too. We're planning a round-robin volleyball match to start at 3 o'clock. I'm supposed to put together one of the teams. But I'm missing a sixth person and can't seem to find anyone." Her voice trailed off again. "Would you be willing?"

Terence stopped again and knitted his brow. He looked at her, and his eyes involuntarily played quickly over her attractive figure. He caught himself, turned toward the chalkboard, and began to rub his chin. "I . . . can't say. I've been really busy lately, and I do need to blow off some steam. I was supposed to go cycling with my girlfriend tomorrow. But we've had some arguments lately, so I think we're probably not going. You say it's a team of six, right?"

"Yeah, six. I didn't know you had a girlfriend!"

"Uh-huh. About a year."

"I guess you're lucky. I'm at a loose end, myself."

The conversation paused. Terence then responded in a carefully measured tone, "You know, it's a little funny to get a request like this, because I'm, well, the TA for a class you're in. I'm a little nervous about how it might look."

"It's just a volleyball game, no big deal," Celia responded hastily. "You shouldn't feel uncomfortable. Bring your girlfriend if you like."

"Leah won't go to this if she won't go cycling," laughed Terence. "Hey, on a different subject—the department head, Professor Peccavi, has the same last name as you. Is there any relation?"

Celia smiled sheepishly. "He's my uncle, although he's almost like a father to me. He always wanted me to major in computer science, since he heads the department. But I like biology better, and that's my major. This class is the only one in his department I'll ever have to take. I want to do well, though. That's why I was so worried last week about my bad quizzes." She paused and eyed Terence. "One thing I like about my uncle is that he always goes out of his way to be nice to my friends—you know, help them out when he can and all that." Celia then hopped up to sit on the proctor's desk she was leaning against, and crossed her legs. "So about the pig roast . . . will you come? The games start at 3, but I'll be there to help with the pig early in the morning."

Terence paused to ponder his answer.

◆ What should he do?

Case 4.2 Items with Dual Work/Personal Use

Terence Nonliquet tapped away intently on the computer in his office, the one he used for his teaching assistant duties in computer science. His girlfriend Leah Nonlibet sat at another desk a few feet away, doing her homework. As Terence worked, he began to hum the theme from *Bridge over the River Kwai*. Leah could take only about five minutes of this. "Terence, you're driving me crazy! Could you stop humming?" she complained testily.

"Oh, sorry. I didn't realize I was doing it," he replied gently.

"Why are you so happy anyway? You've got so much to do these days."

"I'm always happy when I get to try out a new piece of computer software. See this? It's called 'Zipdraw.' It's a killer graphics package."

Leah frowned. "How did you get it? It doesn't sound like something for the class you teach."

"Actually it is. Lower-level Comp Sci courses aren't too densely packed at NTI. I finish the material for my quiz section early lots of times. So I get time to spare that I can use to teach something interesting on my own. I used the course account to buy this package. I showed the students how to use it this week."

Leah leaned forward and squinted to look at the screen. "That doesn't look like classwork. That looks like something for the term paper you've been writing!"

"Oh, it is." Terence responded. "Zipdraw is great for lots of things. I've been using it for some of my lab reports, too."

"Terence, that sounds a little suspicious. I mean, you order software that's not required for the class you teach, and then you use it for your own homework. Does the course instructor know about this?"

"No, I never told Professor Bligh. But he's a low-energy guy. He doesn't care too much about extra stuff for students in his class. As long as no one complains, he's happy. Leah, you're worried about nothing. The use for my class is genuine. I took the initiative to buy the software and set it up for student use. The students really like it. Why shouldn't I get some use for myself, too?"

"Will you continue to have the students use it after this week?" Leah inquired.

"Probably not. I have other things I want to teach them."

"So they get one week of use, while *you* take advantage for as long as you're a TA," Leah declared accusingly.

"Leah, what do you want me to do?"

"Quit acting so shady. Honestly, I wonder about you sometimes."

Terence shed his gentle demeanor. "Come on!" he snapped. "You can take off that holier-than-thou attitude. That graduate student—Bernard—in that lab you work in . . . you knew he was pulling the wool over his advisor's eyes about the mass spectrometer. He really messed it up, and Professor Clark had a right to know. But you didn't say anything!"

"That was different! I wasn't thinking of myself," Leah exclaimed defensively. "Bernard is nice to me, and his equipment problems weren't all his fault!"

"Not so different!" Terence retorted with contempt. "And now you're baby-sitting for Clark's teenage daughters! Talk about sucking up to the boss! He's been treating you like a queen ever since you started! You think that's not at least a little self-interest?"

"Terence, I had good reasons for that, too. Anyway, what I do doesn't affect whether you're acting shady or not. I think you should stop using Zipdraw."

◆ What should Terence do?

Case 4.3 Work Assignment and Supervisor Intentions

"So that's how things are going to change from now on," announced Thorne Mauvais to the evening shift workers in front of him. "As I said, I talked this over with the owner, Mr. Laird, and he approves. We both agree that delegating some moment-to-moment responsibility from me to those next in command will make operations a lot smoother. We expect Pandarus Pizza is going to benefit."

The group broke up and got to work. Celia Peccavi walked to the kitchen with Todd Cuibono. "Hey Todd," she blurted, "you're one of the assistant managers. Now you'll have some things to delegate instead of Thorne. Why do you think they changed things?"

Todd glanced around furtively to make sure no one was listening. "I think Mr. Laird saw how incompetent and lazy Thorne can be," he whispered.

"Well, don't let all the power go to your head!" Celia laughed. "One tin-horn dictator around here is enough!" Todd smiled guardedly. "And make absolutely sure I get no hard jobs!" she continued with a giggle.

Todd's smile turned slightly malicious. "Well, Celia, there's actually been something I've been meaning to tell you. It might as well be now."

Celia did a double-take. "Oh?"

"As you know, Kelly and I are supposed to split the assistant manager duties. We decide jointly which people will do which jobs. We decided that among other things, it'll be your job to clean the bathrooms."

"Clean the bathrooms? That's disgusting work!" Celia exploded. "Someone more junior is supposed to do that, not me!"

Todd could scarcely conceal a grin. "No, now you're supposed to do it."

Celia grew livid. "But . . . last month we all agreed as a group that the most junior person should clean bathrooms. There are three or four people more junior than me! And I talked with Kelly just yesterday about it. She said she thought it was a good policy. I don't believe she agreed to this 'jointly' with you!"

"Calm down," he replied with thinly disguised mockery. "Just because Kelly and I do things jointly doesn't mean we always agree. In fact, this time we didn't."

"So how come you got your way?" Celia demanded.

"Actually, Kelly likes to argue a lot. We disagree quite a bit. Technically, Thorne is supposed to have the final say. But he doesn't have complete freedom. You notice how I was assigned more workers than Kelly, and more of the kitchen."

"More of the kitchen?" Celia broke in.

"Yeah. Like, I control three ovens to her two. Mr. Laird made that assignment."

"So what?"

"Well, around here, the more ovens you control, the more power you have. It's like a sign of wealth. So when Kelly and I argue, Thorne steps in, but he knows the nod should almost always go to me."

"You still haven't told me why I have to clean the bathrooms. There was an agreement!"

"I feel comfortable with the decision," Todd responded airily. "I'm convinced the task matches your skills better than anyone else's."

"You're just mad because I played computer games in Thorne's office a few days ago, but you didn't have the guts," Celia hissed.

"That has nothing to do with it. But you did call me a thief that night. I won't tolerate that any more."

Celia tossed her head. "I didn't call you anything," she sniffed. "I just mentioned some missing food, and asked if you knew where it was. I was just doing my duty."

"And now I'm just doing *my* duty," Todd retorted. "If you don't like my decision, you can take it up with Thorne." Todd paused. "I know you and he are best buddies," he added with dripping sarcasm.

◆ What should Celia do?

CASE 4.4 Disclosing Proprietary Information in Interviews

"Martin, how did your interview go today?" asked Myra Weltschmerz as she salted the pork chop she was cooking on Martin's stove.

Martin looked up and frowned. "Watch what you're doing! You're getting that stuff all over. I just cleaned the apartment a couple of days ago!"

"Sorry. I'll clean it up," she replied timidly.

A brief pause allowed Martin's anger to evaporate. "I don't know. Good and bad."

"What do you mean?"

"Well, the recruiter was running way behind, so we started a half hour late. Then he had to go to some kind of special dinner. Since I was the last one on his schedule, he had to cut me short. I got only about five minutes."

"That sounds all bad. What's the good?"

"Well, he really liked my resume. And ten years ago before he worked for Motorel, he worked for another company that used etchants for some of their processes. I know something about etchants from the stuff I'm doing for Tripos Metal Polish."

"So what happens now?" Myra asked, taking the pork chop out of the pan.

"He set up a special time to see me tomorrow, just before he catches his plane. We'll finish the interview then."

"Awesome!"

"I'm not so sure. Even in five minutes, he started asking me some really detailed questions about what I do at Tripos. So I said I work on their computer system, and go out on some trips for customer relations. But he wanted to know more. He's a chemist by training, so he started asking me about how Tripos makes its acid polishes. I didn't know what to say, because some of that's proprietary!"

Myra's eyes widened. "Do you think he was trying to steal secrets?"

"I doubt it. Motorel doesn't use or make etchants. I think he just wanted to see how broad my knowledge was. I'm a computer science major, right? So he wanted to see if I knew anything else. Plus, I think he was just interested. The trouble is, he'll want to talk about it more tomorrow. It's hard to show him all the stuff I know while talking around the proprietary stuff!"

"But maybe he wants to know so he can pass the ideas on to a brother or friend or something. What are you going to tell him? You can't give away secrets!"

"Don't tell me what I can and can't do!" Martin snapped. "You whined and moaned about that endorsement I was going to do for Paragon Academic Supply, until I got tired of it and skipped the opportunity. I'm not going to let you do that again!"

"Martin, I didn't make you do anything!"

"Yes, you did! You always put on that sad, bleeding-heart face when you want something. Like a couple of weekends ago when we had that fancy dinner date set up, and then you wanted to cancel and go baby-sitting. It was the same thing!" Martin lapsed into a mocking wail of imitation. "'But Dolores needs this interview . . . her whole life depends on me . . . she raises her kids all alone . . .' I had to put my foot down to keep you from standing me up!"

Myra shuddered at the recollection of that incident. Martin had been livid, and it terrified her. "Let's not talk about that any more," she murmured hastily.

Martin settled down. "I still don't know what to tell that recruiter. Maybe I can just tell him to keep the details under his hat. I really doubt it can do any harm."

◆ To what extent should Martin talk about the chemistry he knows?

SUMMARY

This unit has laid important foundations for systematic ethical decision-making. We began by suggesting why ethics is important and how we can discuss it in a rational way. We asserted that a core to ethics exists that is independent of culture or personal opinion, and we limited the scope of ethics to human behavior under conscious control. We saw that codes of ethics can serve useful purposes, but cannot substitute for a more comprehensive approach.

Next, we developed a model for the adult person as a moral creature, consisting of mind, will, and emotions. Classes of good habits exist for each of these, corresponding to the virtues of prudence, justice, and temperance/fortitude. While many aspects of human behavior are habitual, people do not come entirely preprogrammed. We asserted that people should decide and act according to the virtues whenever possible.

We moved on to focus on the exterior dimensions of moral action, and developed a systematic way involving event trees to look at decisions together with their consequences. Finally, we examined the interior dimensions of moral action as described by intention, and laid out a consequence-by-consequence approach to describing intention. We asserted that a fundamental goal of the entire moral life is to continue to progress.

Notice, however, that important elements remain missing from our discussion. We have said nothing about how to weigh good and bad consequences against each other, and have only hinted at how to handle our inability to forecast consequences. Without such methods, we have no systematic way to arrive at a final judgment about the rightness of exterior actions. Furthermore, we have said nothing about how to handle mixed or complex intentions, or how to judge their totality. Finally, we have outlined in only a simplistic way the notion of moral responsibility. Unit 2 will take up all these issues.

Some Words of Caution

This unit has raised some subtle and complicated questions. Situations like those in the example cases crop up surprisingly often in the work place, and do not always yield to easy analysis. Several dangers lurk to trap even those with the highest ideals. On the one hand, we may yield to the tempting simplicity of black-and-white thinking, invoking some set of rules with little thought. Yet not even the wisest set of rules can handle all the complications that arise in everyday living. On the other hand, we may fall prey to rationalization, ignoring in hidden ways the uncomfortable aspects of the situation at hand. Rationalization leads to decisions that suit individual tastes rather than the demands of a mature morality.

Moralists know well the danger of disregarding unpleasant consequences. To highlight the problem, some writers distinguish between the "intention of the action" and the "intention of the person." A historical example will help to illustrate what we mean. Toward the end of the Second World War in Europe, the Allied powers sometimes carpet-bombed entire German cities into oblivion. Allied commanders stated their intention explicitly: to demoralize German civilians. This goal could be called the "intention of the person." However, some people objected that breaking citizen morale was only one consequence of the bombing. Another more immediate consequence, that is, the "intention of the action," was to kill German citizens indiscriminately.[1] These objectors argued that ignoring the importance of this immediate consequence led to wrongly justifying behavior that many people consider immoral under all circumstances—the indiscriminate killing of noncombatants.

No one has created a surefire way to avoid these snares. The truly ethical life carries a lot of risk and requires that we shoulder a large burden of responsibility for our actions. However, as with any other endeavor, recognizing and learning from our ethical mistakes leads to greater experience and wisdom, with a steadily decreasing chance for moral failure.

Note

1. The word "intention" as used in "intention of the action" or "intention of the person" is synonymous in many ways with the word "goal." This book defines intention as a chosen attitude of approval or disapproval, and therefore carries a different shade of meaning.

UNIT TWO

RESOLVING ETHICAL CONFLICTS

"He who will not reason is a bigot; he who cannot
is a fool; and he who dares not is a slave."

WILLIAM DRUMMOND OF HAWTHORNDEN (1585–1649), *ACADEMICAL QUESTIONS*

"Every good man is free."

PHILO (C. 20 B.C.–C. A.D. 45), "NOAH'S WORK AS A PLANTER," BOOK I

5

TOWARD A HIERARCHY OF MORAL VALUES

"Think first, then act; lest foolish be your deed."
PYTHAGORAS (6TH CENTURY B.C.)

On Selecting Principles and Methods

Something very important happens when we select principles and methods for solving ethical problems. After this selection, a specific range of right action appears within the universe of all possible actions. Different sets of principles and methods yield different ranges that often overlap only partially.

The fact that different people often select different sets of principles and methods leads to much disagreement over what is right and what is wrong. There are further complications. Most people can explain most of the moral principles that they believe. (Of course, sometimes people adhere to principles in name only, leading to a severe form of moral failure: hypocrisy!) Curiously, however, many people remain less conscious of their methods. Worse yet, many apply their methods inconsistently—for example, using one method in family contexts and another in the work place. The virtue of justice is particularly vulnerable to this kind of split treatment, as with the father who acts like a dominating tyrant at home but becomes Mr. Equality at the office. In this unit, we seek to improve our ability to identify the principles and methods of moral argumentation. By doing so, we can often reach a more civilized accommodation with one another than would be possible otherwise.

We also seek to improve our ability to arrive at consistent, defensible moral solutions to the complex cases that arise in science and engineering. Interestingly, much debate continues within the discipline of ethics about how to treat moral complexity. Classical virtue ethics in particular has focused mostly on relatively simple situations. However, questions of intellectual property rights, global climate change, genetic engineering, and the like raise exceedingly complicated moral issues. A better approach is needed. While this book does not attempt to answer these questions, it does propose an approach to ethical complexity that appears to be new. Although the basic framework remains rooted in classical virtue theory, the following chapters extend that theory in a way that hopefully will prove useful to practicing scientists and engineers.

Hierarchies of Values: Moral and Nonmoral

Most people agree that some kinds of good are better than others, and that some kinds of bad are worse than others. For example, most agree that deliberately knifing a fellow student ranks worse than stealing a horse. How do we arrive at such a ranking? Is stabbing worse than stealing in all cases? After all, in the American West of the 1800s, a conviction for horse theft carried a sentence of death, whereas a conviction for inflicting bodily injury did not.

Complex situations often present options for action that cause both good and bad things to happen. Choosing the best path requires weighing the goods and bads against each other to determine which way the balance finally tips. Similar questions arise in situations that pit conflicting obligations against each other. Unfortunately, many people rank what is good and what is bad in an unconscious and inconsistent way. As a result, ethical disputes often erupt with little recognition that the real disagreement originates in these implicit rankings. Of course, problems remain even with perfect recognition. Most people agree about the relative importance of some values, but no precise and universal ranking yet exists (and probably never will).

In speaking of good and bad, we must distinguish carefully between moral and nonmoral values. Only the former lie within the realm of ethics. Nonmoral values involve preferences among colors, foods, clothes, music, sports teams, climate, and the like. Moral values involve attitudes toward people and other living things, including loyalty, honesty, humility, arrogance, and the like.

Everyone assigns priorities among nonmoral values. One person might value the fit of a pair of slacks over the color. Another might value price above all. Clearly no nonmoral hierarchy can be universally accepted. Such diversity in ranking should neither surprise nor offend us. After all, Mozart and Michelangelo probably left the world better off by respectively choosing music and sculpture as their most important pursuits.

Everyone assigns priorities among moral values as well. Once again, no universal hierarchy exists. The stakes become far higher in this realm, however, and several problems enter in. The first involves deciding what falls within the domain of morality. In other words, how do we draw the line between moral and nonmoral matters? We need to establish this distinction as best we can before attempting to rank moral values.

Line-drawing

Let's consider an example. Suppose you are hiking on a deserted mountain trail, and suddenly happen upon the edge of a high cliff. For some reason you feel compelled to reach for a small pebble nearby and pitch it over. So you peer over the cliff to check for passersby, toss the pebble, and listen to it shatter on the rocks far below. Suppose further that no local regulation exists to govern rock throwing, and the area has not been set aside for special environmental protection. Does this act fall in the moral realm? Many would respond "no," reasoning that you merely cause a tiny hastening of natural erosion. Now suppose that under the same conditions you spy a large boulder perched at the edge of the cliff. You pry the boulder loose and send it over. Does increasing the size of the rock push the act into the moral arena? How about if you pull up a clump of weeds and toss it over? How about a small ant? A small toad? A baby rabbit? A young fawn? A human infant? A human adult?

Most everyone would agree that tossing a person over a cliff is a moral (or more accurately, immoral!) act. But what about the items between the small pebble and the human adult? We proceeded from inanimate objects to plants to animals to humans. At what point did you say, "There is something wrong with this!"? This example illustrates the concept of "moral importance," the idea that moral stakes are higher in some situations than others. The example also suggests how difficult drawing the line between the moral and the nonmoral can be.

So far we have concentrated on objects—what about actions? Suppose in the example above that instead of walking alone you hike with someone. Suppose that the two of you encounter a narrow part of the trail. There is no foreseeable danger or interesting wildlife on the trail, and no prior agreement about who should lead in a single-file walk. So, because you happen to be ahead at the moment, you take the lead. Many people would agree that this act carries little or no moral importance. Now, suppose you walk so slowly that your companion must watch carefully to keep from bumping into you. Does this action merit higher moral importance? How about if you stop without warning to check your wristwatch? What if you do it several times? Suppose you suddenly stop, turn around, and kiss your companion? What if you then impale your companion with a knife? Most would agree that by this point moral importance has risen very high! For actions as for objects, there exists a continuous line of moral importance, and line-drawing between the moral and the nonmoral presents a troublesome challenge.

Objects and actions are directly observable in some way. What about invisible things like attitudes? Suppose you go to a sports grill with a few acquaintances, who elect you to go to the service counter and order a pitcher of soda to share. Suppose further that your acquaintances express no particular preference of soda. You personally prefer cola, so that is what you order. At this point, your choice remains nonmoral; you have simply picked the highest item in your hierarchy of soda preferences. Suppose now, however, that one of the group, Marty, hates cola but is too shy to express himself openly. You happen to know about his dislike. Further, you don't like Marty because he insulted you during a game of darts the night before. So you order cola with the added purpose of getting even by giving him a drink he doesn't like. At this point, your choice has taken on moral coloring because of your hope to get revenge. Notice that the external action remains the same; the moral content of the choice depends upon your interior intention. Here's where the line-drawing problem comes in. Suppose that Marty has only a weak dislike for cola, and your feelings against him are also weak because he insulted you a year ago rather than the night before. The moral content of your drink order decreases accordingly. At what point does the choice become nonmoral?

In sum, we need to divide a continuous line of moral importance into separate bins labeled "moral" and "nonmoral" for objects, actions, and attitudes. This problem has plagued discussions of ethics for centuries, and we will not attempt to reach a concrete resolution. Unfortunately, leaving the distinction between moral and nonmoral so vague may leave some readers dissatisfied. Accustomed to the precision of mathematics, many scientists or engineers might knit their brows. However, in reality such vagueness crops up in technical disciplines more often than we might suppose. For example, consider the colors in the spectrum of visible light. At what point does red become orange? Or green become blue? While the wavelengths of the spectrum have a precise mathematical description, any attempt to separate the spectrum into discrete colors requires that we make divisions that are ultimately arbitrary.

While some things occupy higher levels of moral importance than others, what determines this hierarchy? Unfortunately, no answer exists that will satisfy everyone. Serious disagreement appears at the higher end of the continuum, for example, where questions of anthropology arise. These issues generate controversy, as the debate over the "animal rights" movement shows well. Some argue that humans have the highest moral importance and should always take priority in situations like toxicology testing. Others quote Ingrid Newkirk's famous phrase,[1] "A rat is a pig is a dog is a boy," and assert that such testing on animals is as wrong as it is on humans. It seems difficult to resolve such disputes without resorting to philosophy or religion—arenas beyond the scope of this book. Hence, we conclude that while hierarchies of moral importance lie near the center of moral thought, different people arrive at different rankings on grounds that lie outside experimental science. While this conclusion does not settle the question, at least it points more clearly to the heart of certain controversies over right and wrong.

An Example

Let's look at some of these ideas in a fictional example.

CASE 5.0 Competition with Coworkers Who Are Friends

"Emily, how come those income figures for last month aren't in the subdirectory I set up for them?" asked Martin Diesirae. He squinted at the screen in front of him, then glanced at Emily Laborvincet, who was also working in the accounting office of Tripos Metal Polish. "You didn't lose them, did you?"

Emily looked up from the pile of papers in front of her. "No, they're not lost. They're in with the figures for the year-to-date."

Martin frowned. "What are they doing there?" he demanded. "I set the system up so the most recent month's figures had their own sub-directory. It's more organized that way."

"Well, it was making my accounting harder to keep them there," Emily retorted. "So I moved them to where they would help me most. It's just as organized as ever, but it makes my job easier."

Martin grew visibly upset. "Who said you could do that?" he cried. "*I'm* the one responsible for setting up the hardware and software for this company. I'm a computer science major!"

"Martin, it's nothing to get excited about. It's not a big deal." Emily answered gently, trying to calm him.

"Yes it *is* a big deal," Martin persisted. "My job is to set up and maintain hardware and software. Yours is to use it for your accounting. I can't do my job if everyone just goes in and starts changing the directory structure when they feel like it. You need to clear changes with me first."

"Martin, I was going to, but you weren't around. You were on one of those customer relations trips with Chase. It's hard to find you sometimes. So I used my own judgment. Monica said I should do that when I need to."

"I don't think this is what Monica had in mind," Martin muttered. "She owns the company. She understands that you can't have every-one doing everyone else's job without telling them!" Martin eyed Emily with faint contempt. "Especially new people."

Emily's voice rose. "This is ridiculous. You're so territorial. So what if I'm new? We've worked well together so far, right?" Emily paused to wait for an answer. "Right?"

"Yeah," Martin admitted grudgingly. "But Monica is grooming me for more responsibilities next semester, when I have hardly any classes and can work here most of the day. Chase's health isn't so good, you know, so he's cutting back his work time. I'll step into his shoes un-til I graduate, so Monica'll have time to find a good replacement for him. I've got ideas for getting our development efforts on-line—Web site, Internet communications, everything!" Martin stood up. "So I don't want anyone trying to take over computer responsibilities. You're re-ally ambitious. You'd be just the one to try," he said accusingly.

"Martin, I am *not* trying to take away your responsibility. Monica says all the time what a good job you do on the computers. But I do have big plans, just like you. I'm only a sophomore, remember. You're a senior. When you're gone, then I might want to step into your shoes."

"Well, I'm not gone yet. Please change the directories back to the way I had them," Martin huffed. Then he walked out the door.

- ◆ What should Emily do about the directory?
- ◆ Have you ever had to compete directly with a friend for job ad-vancement?

Clearly Emily must sort through several issues before she can arrive at a final decision about what to do. To help with this sorting, she might try laying out systematically all the considerations. Then she needs to weigh each one. We can identify at least four things that require balancing:

1. Making her own job easier
2. Advancing her own position within the company
3. Improving the overall efficiency of the company
4. Maintaining good relations with Martin

Items 1 and 2 focus mainly on Emily herself. Whereas at first these items may appear to involve little more than Emily's personal preferences, they fall within the realm of the need for personal growth and happiness. Bal-ancing personal needs with those of others is an important moral concern.

In this case, Emily must choose an action that in effect place items 1 through 3 in opposition to item 4. The relative importance she attaches to the various things will determine how she decides. Any hierarchy of val-ues that does not have item 4 sitting on top will lead her to keep her new organization of the computer directories. Only if she values item 4 more than all the other items combined will she switch the directories back to the way they were.

Mathematical Analogies

We have considered moral importance as a continuous line along which we place objects, actions, and attitudes. We have appealed to physical analogies like the colors of the spectrum. Taken together, these ideas hint that mathematical analogies might prove useful in analyzing ethical matters. Such analogies capture important aspects of moral concepts, and often appeal to those having technical training. Furthermore, in certain limited situations dealing with easily countable items (like distributing money), mathematical equations can find use in practical moral calculations. In fact, we will make significant use of mathematical analogies in the next chapter.

In spite of their considerable usefulness, we need to keep in mind that mathematical analogies in ethics should not be taken too far. Such analogies imply that moral variables can be quantified. Some matters simply do not lend themselves to this kind of treatment. How can we quantify the value of something like a smile? Furthermore, moral variables rarely exist in the abstract; the details of situation-specific circumstances make all the difference.

As we mentioned above in our discussion of line-drawing, the incompleteness of mathematical analogies in ethics may leave some scientists and engineers disappointed. Nevertheless, fuzzy, difficult-to-quantify circumstances are important not only in "subjective" subjects like ethics—they appear even in simpler and thoroughly "objective" situations like a game of chess. Suppose you are ready to capture your opponent's knight in exchange for your bishop. Will you come out ahead? That depends on the hierarchy of importance among chess pieces. Many chess players agree on a numerical ranking for the pieces, assigning one point to a pawn, three to a bishop or knight, five to a rook, and nine to the queen.[2] On such a scale, the exchange appears even. However, experienced chess players know that a knight can outperform a bishop in a cramped position, and can attack targets on squares of both colors. A bishop, however, can vastly outmaneuver a knight in an open position, even though a bishop can reach only half the squares. Also, the power of a bishop greatly multiplies when only a few of the opponent's pawns sit on its color and when its twin bishop remains on the board.[3] To decide whether you come out ahead, you also need to look at exactly what your bishop and your opponent's knight are doing. If your bishop is leading a checkmating attack while your opponent's knight sits lamely out of the action, the exchange inflicts severe damage on your prospects. In some positions a well-placed pawn can prove more valuable than a poorly placed queen! But how do we quantify "well-placed"? We cannot. In short, even in well-defined games like chess, material rankings depend on circumstances. Positions simply cannot be evaluated by a numerical algorithm; nonnumerical, situation-specific considerations play a decisive role. Ethics is far more complicated than chess, so for the same reason we cannot expect to perform numerical analysis in the moral realm.

These points deserve special emphasis in a society that shows an obvious need to quantify everything from soft drink preferences to performance in a job. Some managers seem to demand "performance metrics" for whatever their employees do. To these managers, things that can't be quantified might as well be ignored. The disciplines of science and engineering are particularly vulnerable to this pull toward numerology, because mathematics forms the principal language of technical discussion. Some people claim to "explain" an observation simply by producing an equation to describe it. Those who cannot produce such equations sometimes find that their ideas are not taken seriously.[4]

Ranking the Virtues

Is there a way to rank the classical virtues like justice and temperance, or other moral values like benevolence and loyalty? Such a ranking would help greatly in resolving difficult ethical conflicts. Many moralists through the ages, beginning with Aristotle, have argued that such abstract rankings are hard to make apart from specific cases, and are not terribly useful even when made. For example, one particularly simple hierarchy that originated in the fifth century advises, "Love and do as you please."[5] This hierarchy sets love above all other values, but proves difficult to implement in practice because the word "love" carries so many different meanings. For example, consider the differences between "puppy love," "brotherly love," "romantic love," and "tough love."

In spite of this difficulty, many moralists over the centuries have ranked justice ahead of prudence, temperance, and fortitude in importance. This ordering follows from the way failure to practice these virtues usually comes about. Justice operates within the will, which distills input from the mind and emotions into final judgments. Most moralists attach supreme importance to avoiding malicious evil. Failures in justice often involve malice, while failures in prudence or temperance/fortitude usually involve negligence or cowardice in the mind or emotions. Since malice outranks negligence and cowardice in capacity for wrongdoing, and since the will sits atop the mind and emotions in moral decision-making, the primary importance of justice follows naturally.

Classically, moralists have placed benevolence into a class apart from the natural virtues. While a lack of justice causes definite wrongdoing, a lack of benevolence might not. Thus, there is less obligation to practice benevolence than justice. This means that benevolence lies on a plane higher and more praiseworthy than simple justice (and prudence and temperance/fortitude).

A REAL-LIFE CASE: Scientific Tests Using Animals

Few practices of modern science have caused more controversy than the use of animals in testing. On the one hand, knowledge gained from animal studies has spread into nearly every corner of medical research. Numerous antibiotics and vaccines have been developed with heavy dependence on animal testing. Animals also play a central role in toxicity testing. Organ transplants from animals not only save lives but also provide ways for developing new medical procedures. Given that many experimental tests cannot be performed ethically in humans, in many cases only animals provide physiologies sufficiently close to make such experiments possible.

On the other hand, opponents sometimes claim that animals occupy a moral importance near that of humans, thereby making testing no more ethical in animals than in people. Opponents also highlight the differences in biology among species. For example, it is well known that certain substances that appear safe in common test animals are dangerous in humans, and vice versa. Alternatives to animal tests exist in many cases—autopsies, epidemiological studies, and tissue cultures, for example. Furthermore, sometimes animals are kept in cramped quarters, and some tests inflict significant pain. Toxicity and other tests result in the deaths of large numbers of animals.

There are many shades of gray in this controversy. Animals range from insects to worms to mice to rabbits to dogs. The ability to apply the results to humans varies widely in this progression, with the animals "closer" to humans generally providing more reliable results. However, many people believe that the moral importance of the animals increases in about the same progression. Possible uses of animals range from cosmetics testing to vaccine testing to toxicity testing to organ transplants. Sometimes alternatives to animal tests are readily available, sometimes not. The controversy is unlikely to disappear any time soon.

- ◆ What moral importance would you assign to animals compared with humans?
- ◆ What criteria would you use to judge the acceptability of a given animal test?

References

Barnard, Neal D., and Stephen R. Kaufman. "Animal Research Is Wasteful and Misleading." *Scientific American*, February 1997, 80–82.

Botting, Jack H., and Adrian R. Morrison. "Animal Research Is Vital to Medicine." *Scientific American*, February 1997, 83–85.

Hampson, Judith. "The Secret World of Animal Experiments," *New Scientist* 134 (April 11, 1992):24–27.

"The Master said, 'The superior man thinks of virtue;
the small man thinks of comfort....' "
CONFUCIUS (551–479 B.C.) ANALECTS, BOOK II

Notes

1. For a detailed account of the protest against animal testing, see Ingrid Newkirk, *Free the Animals!* (Chicago: Noble Press, 1992).

2. Yasser Seirawan, *Play Winning Chess: An Introduction to the Moves, Strategies, and Philosophy of Chess from the USA's #1-Ranked Chess Player* (Redmond, Wash.: Tempus Books of Microsoft Press, 1990), 40–41.

3. I. A. Horowitz and Fred Reinfeld, *How to Improve Your Chess* (New York: Collier Books, 1952), 155 ff.

4. Excessive focus on numbers by some people is not new. Numerology lay at the heart of the philosophy of the Pythagoreans over two millennia ago, who said "Everything is numbers." For a discussion on the Pythagorean view of numerology and the Aristotelian response, see Walter Burkert, *Lore and Science in Ancient Pythagoreanism* (Cambridge, Mass.: Harvard University Press, 1972).

5. A famous saying of Augustine of Hippo (354–430). For a detailed discussion of Augustine's thoughts on love, see Hannah Arendt, *Love and St. Augustine* (Chicago: University of Chicago Press, 1996).

Problems

1. Write a page or two describing an ethical dilemma you have encountered in a job you've had where you had to rank some of your values to make a decision. (If you've been lucky enough never to have been confronted with a problem like this, describe one that a friend or relative of yours has had.) Recommend what action you think you (or your friend/relative) should have taken, and indicate what considerations had to be balanced against each other to make that recommendation. Note: you don't have to say what was actually done in real life (unless you want to)!

2. Each case below has a question after it.

   a. Begin by writing down in tabular form the characters and the interests each has in the case.

   b. List any circumstances that have significant relevance to the moral decision at hand.

   c. List the options/suboptions available to the main character who has to make a decision, together with the event tree flowing from each option.

d. List the principal considerations to be balanced against each other in making a decision.

e. Recommend what you think the character should do.

Case 5.1 Peaceful Work Relations versus Efficiency

Celia Peccavi hummed a happy tune as she washed down the sink in the women's bathroom of Pandarus Pizza. She disliked the work, but her mood was so ebullient that even bathroom cleaning couldn't get her down. Her midterms had gone well, and she had won two hundred dollars on a lottery ticket. Also, the volleyball team she organized from her dorm floor had won the round-robin volleyball tournament at last week's pig roast. They won thanks in part to help from her Comp Sci 109 teaching assistant Terence, who had agreed to play for the team. She stopped humming momentarily. "He's cute when he teaches, and even cuter when he spikes the ball!" she giggled to herself.

She finished the sink, and with a sigh of relief bolted out the bathroom door. She nearly collided with Todd Cuibono, the assistant manager, in the hallway. "Sorry!" she snickered.

"I'm glad you have such enthusiasm for your work," Todd responded dryly. "The bathrooms have been spotless. See? I knew the assignment was a good match to your skills!"

"Todd, I'm in such a good mood that not even you can wreck it. I aced my midterms, and won two hundred dollars on a lottery ticket!"

"Good. Happy workers are more efficient."

"You really think so? You know, Todd, I *am* happy today, but it's not because of the way things work around here."

Todd crossed his arms defensively. "What do you mean?"

Celia sensed an opportunity. "Look, Todd—things haven't always been easy between you and me. I'm tired of all this sniping. Let's put it behind us. This place could be a lot happier than it is, and I think people would work better if it was."

"So what do you want me to do?" Todd asked warily. "Do you want some new policy?"

"It's not just policy, it's attitude," Celia continued earnestly. "Like saying please and thank you, hello and good-bye. Like backing each other up when work gets heavy. Right now everyone is measured by whatever they produce, nothing else. There's no incentive to help other people, even when the business would benefit."

"So you want a lot of touchy-feely stuff instead of efficiency."

"Todd, it doesn't have to be either-or," Celia persisted. "I'm just asking for everyone to be respected. And I think atmosphere helps the bottom line, too."

"I don't do atmosphere," Todd declared. "If it's there, it's nice. If it's not, I can't help it. What I want is productivity I can see. My policies aim toward that goal. You forget—I and the other managers have worked here a lot longer than you or anyone else. Any one of us has made more pizzas than all of you put together. What do you know about managing a business? Who are *you* to offer all this free advice?"

"So you won't try to change anything at all?" Celia implored.

"I think we've wasted too much time already. All this discussion keeps us from doing what we're here for—serving customers!" With that, Todd turned and stalked off.

◆ How should Celia proceed?

CASE 5.2 Long Hours versus Personal Development

"I don't know if I can go to the football game with you this weekend, Todd. I'm feeling really crushed for time," Emily Laborvincet moaned to her boyfriend as they walked home from the library. "I'm carrying fifteen hours of courses and working twelve hours a week at Tripos. Plus I'm trying to keep up with my violin practice, which has pretty much fallen through the cracks with midterms."

"Hey, the football game is no big deal. The Peacocks are no good this year anyway. I'll find someone else to go, or I'll just sell the tickets," Todd replied casually.

"Thanks. I need to practice this Beethoven sonata for my violin instructor." Emily stopped walking and squeezed his hand with a smile. "You know, I really like it when you watch out for what I need like this. Sometimes you seem like such a hard-edged guy. It scares me how carefully you calculate what you can get out of a situation. I appreciate that you turned down the vice-presidency of NCS, with all that tainted money, and quit taking food from Pandarus. But I had to rag on you pretty hard. I like it a lot better when you don't have to be nagged to care about other people—when you do it by yourself. You did it a while ago when you told me how to deal with that redneck Floyd at work. I took your advice and spoke to the owner about his calling me a girl. Monica talked to him, and—you were right—he hasn't done it since. Now you're doing it again!"

Todd shifted uncomfortably from one foot to the other. He felt squeamish about this kind of conversation. "Yeah, well I'm glad you like it. I didn't know you were putting so much time into violin."

"I'm trying to get back into it. I played a lot in high school and was first chair for two years. But when I went to college, I stopped practicing so hard, so I'm getting rusty."

"Well, you can't do everything," Todd advised.

"I know, but the violin is important to me. I spend so much time studying technical things as a chemistry major. I need something for the right side of the brain, too! You're in chemical engineering, Todd. That's worse, if anything. What do you do for your nontechnical side?"

Todd raised his eyebrows. "Nontechnical side?" he sniffed with a hint of derision. "I just do what I do. Right now I'm a student and I'm working. It's going fine. What else is there to worry about?"

"Lots!" Emily scolded firmly. "Part of the goal of life is to become a complete, well-rounded person. We have to learn to think with our intuition, not just our rational minds! We need to think logically *and* to appreciate beauty. These capacities don't grow by themselves. They need to be nurtured consciously."

Todd's voice assumed a mocking tone. "Well, you go ahead and nurture your intuition. I'm going to get ahead in life."

"Is that all you think about?"

"What's wrong with trying to get ahead? Let me tell you, it's a dog-eat-dog world out there. There are plenty of people willing to let you do something for them without giving anything in return. Just like where I work . . . you remember how they wanted me to set up a fancy new computer system, for the same lousy wage they pay any other worker? They told me how great a job I would do, how important it was to the firm, and all that. But they wouldn't cough up any extra money, so I refused." Todd paused and scowled at Emily. "With so many people in the world working against your getting ahead, why work against yourself?"

Emily's fervor began to wither in the face of Todd's disapproval. "I *am* getting ahead with my violin playing. You heard me play last Monday night. Didn't you think it was good?"

"It sounded OK," Todd admitted reluctantly, "but I'm no judge of that. How do you really know you're any good?"

"My violin instructor said so," Emily retorted triumphantly. "Two weeks ago. She said it won't be long before I'll be able to play in serious competitions."

"So who is this instructor?" Todd parried. "What does she know?"

"She's very well known around here. She's in the music school at Penseroso."

"How nice. Is she famous? Has she won any big awards? Is she in the National Academy of Violin Players, or whatever they call it?"

"I don't know, Todd, I haven't seen her resume. What difference does it make? Do you need to be in the 'National Academy' to have an ear for music? I don't think ability to judge good music has anything to do with how many awards you've won!"

"Hah!" Todd snorted. "Look, if it makes you feel better, I used to play the piano years ago. I was pretty good. But now I have other things to do. . . ."

"Like get ahead in life?" Emily burst in. "Todd, I shared this with you because it's hard to have so many exciting things to do and not have the time to do them. I'm playing the violin seriously this semester, but it's affecting my grades. The way things have been going, I may lose half a grade point from my semester average unless I cut way down on this practicing. I get headaches more than I ever used to, and I don't sleep as well. I'm going to have to make some choices about how many hours I will spend on school, how many on the job, how many on music, and how many on the rest of my life. I was hoping to get more from you than snide comments and junk about the National Academy of whatever!"

"Well, I guess I'm lucky I don't have your problem. I know what I'm doing. My grades are good, and that's what's going to get me a job when I graduate. That's all I can say," Todd huffed. "Let's not talk about it any more."

◆ What should Emily do?

CASE 5.3 Long Hours versus Good Personal Relationships

"Hey Terence, will you be able to come to that concert I asked you about last week? You said you'd tell me by yesterday."

Terence Nonliquet pulled his head out of the book in which he had buried it. He slapped his hand to his forehead in disgust, and began to apologize to his girlfriend Leah Nonlibet. "I'm sorry, Leah," he groaned. "I forgot all about it. No, I won't be able to come. Just yesterday I agreed to visit my mom that weekend."

Leah looked flabbergasted. "You forgot? I've been reminding you for weeks! And visit your mother? You're on the phone with her all the time! And she just visited here a couple of weeks ago!"

"Leah, I'm sorry. But I promised. I have to visit her. Isn't there something else we can do together?"

"Terence, no. I wanted to go to this concert. Not just any concert, *this* one. Why don't you call your mother and tell her something came up, and you'll visit her another time?"

Terence shook his head. "You know how my mother is. She takes stuff like that very personally. I can't make promises to her and then break them right away!"

"Well, I take stuff very personally too, you know," Leah huffed. "And you promised me you would go once you checked your calendar for conflicts. You never checked. I consider that a broken promise. How come you can break promises to me but not your mother?"

"Leah, it's not like that," Terence whined.

"Oh yes it is," Leah broke in. "And I'm tired of it. A girl has a right to some time from her boyfriend, you know. I have a lot of friends with boyfriends. And their boyfriends spend a lot more time with them than you do with me!"

"I've been busy! I have a bunch of hard classes, and I'm a TA. My grades are good, and I'm a popular TA. That doesn't come free—it takes time. I try to spend time on you, too!"

"It's not that simple," Leah persisted. "You spend extra time with some of your slower students, beyond what normal TA duties require. You said so yourself. Even your mother thinks it's too much. If you want our relationship to grow, you have to spend more time on me."

Terence's face darkened. "I don't think you're being fair," he exclaimed. "We study together all the time. We go on bike rides. We talk on the phone every night if we don't see each other. I admit, this past week or two I've gotten distracted, but overall I think I do pretty well . . . considering our busy schedules."

"But I don't feel like it's enough, and my friends don't either."

Terence's voice started to rise with anger. "I don't like it when you tell your friends our troubles! It's none of their business! And what do they know? Take Mia. She's an emotional black hole! She goes through one boyfriend after the next, suffocating every one until they dump her just to get some air. And then she insists they're all jerks. Lots of your friends are that way." Terence leaned forward and glared at Leah in silence for several seconds. "And you're getting to be that way too," he fumed.

Leah felt her stomach sink away. She tried to conceal her anxiety, without success. "I don't know what you mean," she quavered.

"You demand too much," Terence continued relentlessly. "Nothing I do satisfies you. You just need more, more, more."

Tears welled up in Leah's eyes. Terence sensed that to press further would invite a major, maybe irreparable, breach. He backed off, and forced his voice to soften. "Leah, there are only twenty-four hours in a day. I want to study, I want to teach, and I want you. I want to find a way, but there have to be priorities. Let's talk about it another time."

♦ What should Terence do?

Case 5.4 Geography, Timing, and Career

"Guess what, Myra? Motorel offered me a site visit!" Martin Diesirae cried jubilantly to his girlfriend Myra Weltschmerz. He wrapped her in a great bear hug. "They just called me!"

"Great," she gasped, dazed by the hug. Myra was not used to such a warm reception upon walking into Martin's apartment.

"It sounds like a great opportunity. It's in East Cupcake, Idaho."

Myra tried to smile, but it wasn't easy when she could hardly breathe. Martin finally released her. "Motorel . . ." she responded, trying to catch her breath. "Isn't that the company that wanted to know about all this proprietary stuff you're doing now at Tripos?"

"Yeah," said Martin, calming down.

"You didn't tell them anything, did you?" Myra asked reproachfully.

"Nah, I didn't have to. Actually, the recruiter was impressed when I said I couldn't tell him everything I know about. He understood my problem right away, and said he respected people who could keep important secrets. He laughed and said he'd find other ways to gauge my knowledge. I guess he did!"

Myra's face brightened briefly. "That's good!" she agreed. But her face rapidly turned somber. "Martin, when would they want you to start if they hire you?"

"Next spring, right after my graduation."

"But Martin, I thought you said last week that you were thinking of delaying graduation until next December. Remember? You said Tripos is in big trouble right now because the owner and customer relations guy both got hurt bad in a car crash. They can't work, so you were maybe going to work full-time next semester to help the company through till they got better."

Martin's brow furrowed as the realization sank in. "Oh, yeah. I was so happy, I forgot. Not too many people are getting site visits this year, you know. I like Tripos, and they really need me right now. But there's no way I could put Motorel off until a year from next month!"

"And what about your family?" Myra continued. "Your mother is sick, and may not live long. You have to help her out on weekends sometimes when your dad is out of town. You can't do that from Idaho!"

"Well, I'll just . . . I can . . . I don't know!" Martin sputtered with exasperation. "You sure know how to throw cold water on things!"

Myra's voice began to shake. "And what about me? I'm graduating in spring, too. You told me East Cupcake is a little town in the middle of nowhere. How am I going to get a job?"

"Maybe Motorel needs environmental engineers like you," Martin remarked offhandedly.

"Martin . . . I didn't tell you this, but I tried to interview with Motorel, too," Myra confided nervously. "But they said they weren't looking for someone like me. So you see? I can't work there."

"Maybe they've got some kind of employment program for spouses or significant others," Martin suggested. "Anyway, there must be some kind of job you can get in East Cupcake."

"Like what, wait tables? Martin, I didn't spend four years in environmental engineering to take some dead-end job I don't like."

"So what am I supposed to do? Cancel the trip?" Martin snapped. "The market is tight. Who knows if I can get another good one!"

Tears suddenly rolled down Myra's cheeks. She began to cry. "I don't know. But we never even talk about how we can stay together after graduation. I think about it all the time. . . ."

Martin softened. "I'm sure we'll work something out," he offered uneasily. He embraced her again.

"We won't if you take trips like this," Myra sobbed. "You're like my dad all over. After the divorce, he promised I could visit all the time. But he had a girlfriend in another state. So a few months later he moved away. Then he promised to send me plane tickets four times a year. But he was always broke. So I was lucky to get them—maybe once a year." She looked up at Martin through a flood of tears. "Now you're making promises, too," she whimpered.

◆ Should Martin go on the site visit?

6

STARTING MORAL JUDGMENTS: EVALUATING EXTERIOR ACTS

"Men decide many more problems by hate, love, lust, rage, sorrow, joy, hope, fear, illusion, or some similar emotion, than by reason or authority or any legal standards, or legal precedents, or law."

Marcus Tullius Cicero (106–43 b.c.), *De Oratore*, ii, 178

Moral action has both exterior and interior aspects. The exterior dimension concerns actions viewed from outside the person, whereas the interior dimension concerns personal intention. In this chapter we focus on the exterior dimension and develop a systematic way for analyzing it in complex moral situations. *Note that any judgment of moral goodness based only on exterior actions and consequences remains incomplete. Intention plays a crucial role in the final evaluation.* The next chapter deals with this aspect of ethical judgment.

A Mathematical Analogy

In complex cases, an action may lead to many possible consequences, some good and some bad. Some of these consequences might be far more important than others, and some might be far more likely than others. To avoid confusion, we want to have a systematic method of listing and balancing all these consequences in a way that includes how important and how likely they are. The method we develop here draws on a mathematical analogy, and has the very important advantage of being straightforward and systematic. Furthermore, our method resembles the way engineering risk-benefit analysis is sometimes done,[1] and therefore should appeal to the mindset of scientists and engineers. Interestingly, the method appears to be new to classical ethics. (See note 2 for a discussion of how this approach tries to combine the insights of classical virtue theory with the clarity and precision of the "utilitarian" method.)

Many people may remain suspicious of any attempt to reduce a complex subject like ethics to mathematics. This suspicion is well founded. There is no good way to quantify some moral "variables." How can we put a number on something like loyalty? Thus, we should remember that our mathematical analogy is *only* an analogy. It should *not* be taken literally. For this reason, we will carefully avoid putting actual numbers into any "equation," and will not read too much into functional form. In this way we should be able to bring some of the systematic clarity of mathematics into ethics without stretching the ideas too far.

With this caution firmly in mind, let's consider what we have to do to balance all the good and bad that can follow from an action in a complex situation. For each thing that happens, we need to decide first whether it is good or bad. Then we need to decide how important it is and how likely it is, since consequences that are more serious or more likely should tip the balance more strongly. Once we've made these decisions for each consequence, we can add them all together to gauge the overall balance of good and bad that springs from the action. A mathematical analogy then looks like this:

$$\text{Net goodness} = \Sigma \, (\text{goodness of each consequence}) \times (\text{importance}) \times (\text{likelihood})$$

$$(6.1)$$

In Eq. 6.1 the symbol Σ has its usual mathematical meaning of "summation" over all the consequences.

The factor describing the goodness of each consequence describes how that consequence squares with the virtues. We can rate consequences that reflect justice, prudence, temperance and fortitude as "good," and those that do not as "bad." When doing this rating, we can often clarify our thinking by identifying the virtues with which the consequence connects.

The factor describing the importance of each consequence determines *how* good or bad the thing is. Good measures to use for importance might be "high," "moderate," "low," or "zero." Making this evaluation depends on which moral values we think are most important (that is, on our hierarchy of moral values as discussed in the last chapter). For example, the layoff of an unneeded clerk will matter a lot to the clerk but much less to the personnel officer in a remote office who signs the needed papers. In making this assessment, we should try as best we can to take the position of an impartial observer viewing the situation from the outside.

The factor describing likelihood can also be rated by words like "high," "moderate," "low," or "zero."[3] It's important to remember that the likelihood of a consequence represents a *cumulative* probability, incorporating the likelihood of all events leading up to that consequence. Thus, if an action leads to an unlikely consequence A, which in turn almost certainly leads to consequence B, the cumulative probability of event B is still low.

Consider, for example, a commercial jet pilot whose plane is at the end of the runway ready to take off. The pilot is making final flight checks and deciding whether to actually take off. One possible consequence of taking off is that the plane crashes, catches fire, and burns the passengers. If the plane does crash, the likelihood of injury from the resulting fire is fairly high. However, suppose the weather is good and the plane has been reliable, so that the chance of the plane crashing is very low. The *cumulative* probability of passenger injury from fire therefore remains very low. Thus, the entry for likelihood in Eq. 6.1 for the event "injury from fire" should be "very low."

An Example

Let's see how these ideas work in practice by considering an example case.

CASE 6.0 Giving Company Goods to Friends

The clock ticked grudgingly past 9 p.m. on Wednesday night. The Pandarus Pizza Parlor momentarily was empty of customers, and Celia Peccavi busied herself with cleaning the cash register area. Suddenly a rowdy group of four young women burst in—dormmates from her floor.

"Hi Celia! What's cookin'?" giggled one. "Wait, don't tell me! Is it . . . pizza?" The newcomers dissolved in uproarious laughter.

Celia smiled weakly at the bad joke. Then she straightened up and crossed her arms. "What are you guys doing here? Shouldn't you be studying or something?" she demanded with obviously pretended sternness.

"We just came by to say 'hi,'" added another. "Say, it smells good in here. And it's empty, too. How much stuff do you have left over?"

Celia instantly grew more nervous. This was not the first time her friends had come around looking for a free handout. She glanced quickly behind her, but all the other workers were in back, apparently out of earshot. "Shhh!" she whispered. "Do you want to get me in trouble?"

"Of course not! But we have a major case of munchies, and thought we'd visit to take some excess food off your hands. You let us have some once before!"

Celia continued to look about. "That was one time, when we had a lot of miscooked orders. That food was going in the garbage anyway. We usually don't make so many mistakes. There's no extra tonight."

"Oh, come on, Celia . . . we still have the munchies. Can't you come up with at least a few breadsticks or something? Who'll know? And remember when Aida here picked you up last week on a moment's notice when your car ran out of gas? She sure saved your butt! Don't you think you can show a little appreciation?"

Celia hesitated, and glanced around again. Her lips tightened for a moment. Then she whispered, "OK, just some breadsticks. But get them outta here fast. I don't want to get caught."

It was all over in fifteen seconds, while Celia's friends high-fived each other with muffled giggles. The bag of food changed hands, and in a flash the restaurant was quiet and empty again. Celia breathed a sigh of relief.

The sigh wedged awkwardly in her throat, though, when Todd Cuibono suddenly materialized from the back work area. He eyed her carefully. A pregnant pause ensued, and Celia flushed ever so slightly. He launched a casual but calculated gambit: "I thought I heard some customers."

Celia tried mightily to hide her nerves. She leaned casually against the serving counter and nodded toward the restaurant entrance. "Yeah, a few friends from my dorm stopped by. A real bunch of clowns!"

Another pause followed. "Did they buy anything?" he ventured.

"Not much . . . just a few breadsticks," she responded, glancing at the now-empty bin she had taken them from.

Yet another pause. "Really?" Todd's tone grew perceptibly accusatory. "I don't remember hearing the beeps of the cash register."

Celia flushed further and looked down. "Well, my friends were loud, and sound doesn't always carry in the back. Maybe you just missed it." Their eyes met after yet another pause. Todd looked at the cash register, then back at her. He stroked his chin studiously, still gazing at her. Her face now registered fear. They both knew she'd been snared.

After a seemingly endless pause, Todd's face hardened. "Hmm . . . ," he grunted. "You know, I've told you before, it's my job to see that things run smoothly around here," he continued gravely. I suggest you keep your 'clown' friends out of here and get back to work." He turned away abruptly and left Celia standing alone at the counter while he pondered what to do.

♦ Was it right for Celia to give the breadsticks to her friends? Why or why not?

♦ If not, what should she have done? Why?

♦ Was it right for Celia to lie to Todd? Why or why not?

♦ If not, what should she have done? Why?

♦ Should Todd pursue the matter further with his own boss?

This example actually involves two separate choices: Celia's decision whether to give the breadsticks to her friends and her decision whether to lie to Todd. Let's examine each of these in turn.

1. Celia's Decision Whether to Give the Breadsticks

Figure 6.1 diagrams the events that flow from two possible choices: giving the breadsticks to her friends or withholding them. Both "event trees" take fairly simple forms.[4] Notice that for simplicity the event trees for these opposite decisions do not include exactly opposite consequences. We omit those that would happen anyway even if this situation had never occurred— Pandarus not losing profit, for example, or Celia not settling her score with Aida. However, disappointing Celia's friends should appear under withholding the breadsticks because if the friends had not entered the restaurant, they would have become neither happy nor disappointed.

If Celia gives the breadsticks, she settles a debt of sorts with Aida. Celia makes her friends happy and may even strengthen her friendship with them. However, Pandarus Pizza loses a few dollars of profit, and Celia breaks a strict company rule. Let's consider the goodness of each consequence first. In general, it's good to settle debts in accord with fairness. It's also usually in the best interests of people to make them happy, again in accord with fairness. Whether strengthening Celia's relations with her friends is good or bad depends on what kind of people they are. Since we know little about the visiting women, however, common experience suggests that stronger friendships work to everyone's best interests. On the other hand, losing profit is not in the best interest of a business. Moreover, company regulations normally seek both equity toward employees and the best interest of the organization. Since the company rule against

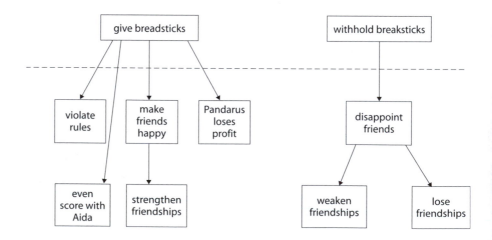

FIGURE *6.1 Event trees for Celia's decision to give breadsticks to her friends*

giving away food does not seem unjust, breaking that rule offends against fairness.

Let's consider importance next. Some consequences seem to have at best low or moderate importance. The debt to Aida is not formal; the increased happiness of Celia's friends is neither large nor lasting; and Pandarus loses only a few dollars. On the other hand, the strictness of the rules suggests great importance.

Finally, let's consider likelihood. All of these consequences, except possibly strengthening the friendships, seem quite probable.

We can now evaluate Eq. 6.1 for this option by listing each consequence together with the most relevant virtue. Below each consequence we can show the goodness, importance, and likelihood. A convenient way for arranging these items appears below:

even score make friends happy strengthen friendships

(fairness) (fairness) (fairness)

Net goodness = (good)(low)(very high) + (good)(low)(very high) + (good)(moderate)(moderate)

Pandarus loses profit violate rules

(fairness) (fairness)

+ (bad)(low)(very high) + (bad)(high)(very high)

Crudely speaking, the first and fourth terms cancel, leaving the second, third, and fifth. The second and third are good and have roughly the same weight, since the higher importance of the third term is compensated by its lower probability. The fifth term is bad with heavy weight, however. So the sum comes out not far from a net zero.

If Celia withholds the breadsticks, she disappoints her friends and might weaken her friendship with them or even lose them as friends. Disappointing friends usually offends against fairness, as does weakening or losing friendships (assuming that the friends are worth having). We can therefore rate all these consequences as bad. However, there is no evidence that Celia is terribly close to these friends, so the importance of temporary disappointment and weakening the friendships remains low. The importance of losing them completely might be a bit higher, but the chance of it happening seems lower.

Thus, for not giving away breadsticks, Eq. 6.1 takes the form:

disappoint friends weaken friendships lose friendships

(fairness) (fairness) (fairness)

exterior goodness = (bad)(low)(very high) + (bad)(low)(moderate) + (bad)(moderate)(low)

This time all three terms are bad. Crudely speaking, the sum seems to come out modestly bad.

Thus from a purely exterior perspective, it seems better for Celia to hand over the breadsticks. More generally, it's not surprising that small-scale looting from employers occurs commonly. *However, our ethical analysis is not complete; we have completely ignored intention.* A final ethical judgment requires us to include intention, which we will do in the next chapter.

2. Celia's Decision Whether to Lie to Todd

Figure 6.2 shows event trees for two possible options: to tell the truth or to lie. If Celia tells the truth, Todd could tell his boss, who in turn could at least cause trouble for Celia or even fire her. Firing would discourage other employees from giving away food, and Celia could have trouble paying her bills if she cannot rapidly replace the lost income.

We will not explain every detail of the event tree point by point, and reasonable people might disagree about some of the entries. However, as the trees are drawn, Eq. 6.1 takes the form:

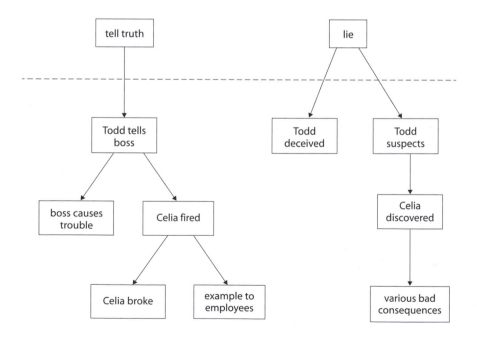

FIGURE 6.2 *Event trees for Celia's decision to lie to Todd*

Todd tells boss	boss causes trouble	Celia fired	Celia broke
(?)	(fairness)	(fairness)	(fairness)

Net goodness = (?)(zero)(very high) + (bad)(moderate)(high) + (bad)(high)(low) + (bad)(high)(low)

example to employees
(truth, fairness)

+ (good)(moderate)(low)

By itself, Todd's telling his boss has no real importance as evidenced by the question marks for that entry. It's the boss's possible anger that really matters. The entries shown above rate the anger together with Celia's possible firing and financial problems as bad for Celia. This evaluation can be debated, since Celia might be inspired to change her ways (a good thing). Celia's firing would probably discourage fellow employees from giving away food, but the discouragement is cumulatively unlikely because Celia's firing is unlikely. In the final summation, the first term drops out, and the last term only slightly balances the remaining three. The sum is therefore modestly bad.

If Celia lies and is only suspected by Todd, little else happens. The story hints that Todd already dislikes Celia, so his relationship with her will probably not change much. Hence, no other consequences show up in the event tree. However, if her lie is discovered, several other things could happen. If Todd tells his boss, then many bad things happen. Equation 6.1 thus takes the following form if Celia chooses to lie:

Todd deceived	Todd suspects	Celia discovered
(truth)	(truth)	(?)

Net goodness = (?)(zero)(moderate) + (bad)(very low)(high) + (?)(zero)(very low)

several mostly bad consequences
of very low probability

+ (bad)(low to high)(very low)

Again, question marks appear where a consequence has no real moral importance. The overall expression sums to slightly bad.

Once again, from a purely exterior perspective, it seems better for Celia to lie. More generally, it's not surprising that lying is common. *However, once again our ethical analysis is not complete; we have completely ignored intention.* A final ethical judgment requires us to include intention, as we will do in the next chapter.

A REAL-LIFE CASE: Chemical Disaster at Bhopal

For nearly two hours on the morning of December 3, 1984, 40 tons of methyl isocyanate (MIC) poured from a storage tank at a Union Carbide pesticide plant in Bhopal, India. Plant workers were unaware of the problem until their eyes began to burn. Even then, no one tried to correct the problem or inform those who lived in the densely populated area surrounding the plant. A huge, lethal cloud of MIC drifted over Bhopal. By the time the episode had ended, over 3500 people died and tens of thousands more were injured. Immediately Union Carbide claimed "full moral responsibility" and gave out twenty million dollars in disaster aid. Shortly thereafter, however, the Indian government sued for three billion dollars, and a massive legal tangle followed that delayed any more payments. In 1989, Union Carbide and the Indian government tentatively settled for $470 million. However, India then sued for more, creating another legal snarl that remains incompletely resolved to this day. It took until 1993, almost ten years after the accident, for even small amounts of settlement money to begin trickling to the victims.

The period before the disaster saw conditions at the Carbide plant that had all the prerequisites for a serious accident. The plant was located in a densely populated area, but no evacuation plan had ever been devised. In fact, many key local officials did not know of the dangers posed by the plant. Those who did know and protested were overruled by the state government, reportedly to protect the highly paying jobs provided by the plant, especially for former government officials. Government-sponsored safely inspections were lax. Union Carbide itself did not oversee the plant carefully, partly because Carbide owned only 51 percent. The other 49 percent was owned by Union Carbide India Limited and the Indian public. The plant was originally staffed by a sizable fraction of highly trained workers from the United States. However, financial pressures on Union Carbide as well as the desire of the Indian workers for more control resulted in the withdrawal of the U.S. workers, leaving poorly trained local management in charge. Production and maintenance personnel failed to communicate; safety meetings dropped off; and staffing was reduced below safe levels. By the time of the accident, numerous valves and pressure gauges were leaky or broken. Key refrigeration and scrubbing units were either under repair or simply switched off.

During the litigation following the accident, Union Carbide claimed that an angry employee had let water into the tank as an act of sabotage. While such a scenario is possible, Carbide never convincingly proved its case. Leaky valves could also have caused the problems. In any case, the technical and organizational problems of the plant made it vulnerable to any unforeseen occurrence.

- ◆ What examples can you give from your own experience of "accidents waiting to happen"?
- ◆ Which deficiencies in the Carbide plant offered the greatest likelihood for trouble of some kind?
- ◆ Which deficiencies offered the most potential for catastrophic (as opposed to minor) consequences?

References

Kurzman, Dan. *A Killing Wind: Inside Union Carbide and the Bhopal Catastrophe.* New York: McGraw-Hill, 1987.

Shrivastava, Paul. *Bhopal: Anatomy of a Crisis.* Cambridge: Ballinger, 1987.

"For what is liberty than the unhampered translation of will into act."
DANTE ALIGHIERI (1265–1321) *LETTERS*, 6, 1311

Notes

1. See Chapter 14 for a further discussion of risk-benefit analysis.

2. Some readers familiar with philosophical ethics may view Eq. 6.1 as closer to utilitarianism than to virtue theory. In fact, the approach described in this book attempts to reconcile the two methods. Utilitarianism employs the principle of utility: that actions should be chosen that lead to consequences having the greatest total balance of benefits over harms for all concerned. There are several ways to calculate this balance: by acts and by rules, for example. However, benefits and harms are often evaluated according to some largely *nonmoral* standard like human pleasure. Furthermore, only the total sum matters, not the individual terms in the summation. Thus, in principle even very unfair actions can be tolerated as long as the total sum of benefits over harms remains large. For example, a small minority could be ruthlessly enslaved for the benefit of a large majority.

Equation 6.1 also balances the results of consequences, thereby imitating the very methodical approach of utilitarianism. However, Eq. 6.1 deviates from utilitarianism in one crucial way: by evaluating benefits and harms explicitly according to the *moral* standards of the virtues. The goodness of each consequence is determined entirely by how it squares with the virtues. For example, unfair distribution benefits or harms would show up explicitly as a significant negative consequence.

The approach of this book differs greatly from utilitarianism in yet another way that will become more obvious in the next chapter. In contrast to virtue theory, standard utilitarianism largely ignores interior morality and character formation. Chapter 7 explicitly treats these issues as a key element in judging the goodness of moral decisions, but again attempts to imitate the clarity and precision of utilitarianism by discussing character formation in terms of the interior consequences of moral decisions.

Finally, Eq. 6.1 moves beyond classical virtue theory (and to a lesser extent classical forms of utilitarianism) by including the idea of likelihood explicitly and systematically.

3. Although we are employing the word "probability," in fact there is a subtle problem with this. Probability presupposes a large number of identical events on which one can perform statistical analysis. No such large number exists for most of the events we are considering. The people, places, times, and circumstances vary from case to case in ways that usually make rigorous statistical analysis difficult, if not impossible.

4. Event trees have been introduced formally in Chapter 3. However, the example shown here should be sufficient to show how they work if you have not read that chapter.

Problems

1. Write a paragraph or two describing an ethical dilemma you have encountered in a job you've had. (If you've been lucky enough never to have been confronted with a problem like this, describe one that a friend or relative of yours has had.) Analyze the case as follows:

   a. List the options/suboptions available to the person who had to make a decision, together with the event tree flowing from each option. Indicate your estimate of the probability for each consequence on the tree (high, moderate, low).

   b. Write down an expression for exterior goodness according to Eq. 6.1 as shown in the text example.

   c. Recommend what you think the person should have done. Note: you don't have to say what was actually done in real life (unless you want to)!

2. Each case below has a question after it.

   a. List the options/suboptions available to the main character who has to make a decision, together with the event tree flowing from each option. Indicate your estimate of the probability for each consequence on the tree (high, moderate, low).

   b. Write down an expression for exterior goodness according to Eq. 6.1 as shown in the text example.

   c. Recommend what you think the character should do.

Case 6.1 Whistleblowing on Safety Violations

"Todd, did you see what Kelly just did?" whispered Celia hoarsely in the corner of the kitchen at Pandarus Pizza.

Todd looked up from the pile of orders he was reviewing. "No, what?"

"She spilled a whole bag of Italian sausages all over the floor, and then just picked them up and put them back in the bag! The chef is putting them on pizzas now!"

Todd raised his eyebrows. "Are you sure?"

"Yes, yes!" Celia blurted out loud. She could hardly contain herself. Then she dropped her voice again. "Didn't you hear the thud when they fell?"

"Oh yeah, so that's what it was. . . . Does Thorne know?"

"That's the worst part! He saw the whole thing and said nothing! And he's the manager! This has got to be, like, massively illegal!"

"Relax, Celia. Don't get so excited. Did you ask Thorne about it? Maybe there's something you don't know."

"What's not to know? You're as bad as he is!" she sputtered.

"No, I'm concerned too. We just need to be sure we have facts. Why don't you talk to him? I'll wait here and listen."

Celia whirled around and tramped right up to Thorne, her fists clenched and her eyes flashing. "How can you take those filthy sausages and put them on people's pizzas?" she shrieked.

Thorne bolted upright and recoiled slightly. "I didn't put anything on anyone's pizza," he contended. "I'm not even cooking!"

"But you saw them fall on the floor, and then let the chef put them on!" she persisted. "You're the manager! You're responsible!"

Thorne glanced around quickly. Work had halted, and everyone was staring at him. His face reddened. "Look, I don't know what you're talking about," he snapped. "How do you know whether I saw something fall or not?"

Celia turned purple with rage. "Oh, so you're going to lie on top, huh?" she shot back. "I can't believe this! It's not the first time, either! I've seen you! Workers come out of the bathroom without washing their hands! You see it and say nothing! Last week you told the chef to use Canadian bacon that was way past expiration! It's a wonder people don't fall over dead after eating here! And it's all your fault!" Celia's eyes suddenly began to well up with tears, and she fled to the bathroom.

Silence followed. After a few seconds, Thorne looked around and laughed weakly. "She's a lunatic, you know." Then more forcefully, "OK, everyone back to work!"

Ten minutes later, Celia emerged from the bathroom more composed. Todd spied her and quickly sought her out. "Are you OK?" he ventured. She nodded. "That wasn't what I had in mind when I suggested you talk to him," he continued.

"But he's such a jerk! I just can't stand him. He does stuff like this all the time. A while back he wanted me to make pizzas faster by undercooking them. I was going to tell the owner, but Thorne threatened to fire me if I did. I didn't want to go looking for another job right then, so I stayed quiet." Celia paused for a moment, and smiled maliciously. "But at least I got away with not undercooking them that night . . . he got distracted by some angry customers." Celia paused again, then continued, "And do you know what else he does? He takes bribes. I saw it happen. A window painter gave him some football tickets in return for Pandarus business. I'm tired of his getting away with all this. This time I'm going to report him, if not to the owner, then to the health inspectors."

"That might not be the best way," Todd cautioned.

"I don't think you care about the law, either!" she chided harshly.

"I'm only thinking in the long term, Celia." His voice dropped. "I don't like Thorne, either. I want to see him out. You and I can work together. We need hard evidence for the inspectors, and he can simply deny seeing anything. The owner might not force him out, and then Thorne'll still be kingpin. But he knows we're onto him. His position is weak. He can be persuaded to give up some of his power, so that someone else can stop the abuses."

"Give up power? To whom?" Celia asked suspiciously. "To you? Maybe you're trying to maneuver for his job!"

"I'm not thinking about myself. I'm just. . . . "

"*That* would be a first!" Celia broke in derisively.

"I'm not thinking about myself." Todd repeated tensely. "I'm just trying to find a better approach than screaming my head off like you did."

"I still think he should be reported, regardless. He's breaking the law."

◆ What should Celia and Todd do?

Case 6.2 Hiding Convictions on Job Applications

"Mmm, the lasagna smells great!" exclaimed Martin Diesirae to his girlfriend Myra Weltschmerz. "You work miracles with my apartment's oven! Do you cast a spell on it?"

Myra grinned broadly and basked in the compliment. "I made your favorite recipe," she responded.

"Well, it won't last long, 'cause I'm starved. All the extra work at Tripos builds an appetite."

"How's that going?"

"Monica, the owner, and Chase, the sales manager, are still in the hospital from the car crash. It looks like they'll be in for a while, yet. At least the cops found the bozo who hit them. He was intoxicated and drove away from the scene. But some friend of his found out and turned him in."

"But Monica and Chase are like the brains of the company!"

"Yeah. There are a few chemists who do product development

and quality control. But the ones we have happen to be short on business and management skills. They're just too nerdy. So we're muddling through as best we can. I talked to Monica on the phone yesterday for a few minutes. She sounded really bad. But she said Emily Laborvincet, a part-time student from Penseroso, should take over all accounting and the computer work I did. She wants me to take over Chase's job, and serve as general-purpose troubleshooter."

Myra put the steaming lasagna on the table between them. "That's a lot of responsibility," she said. "I've been working hard, too. I had two job interviews today, and I sent out a bunch more resumes." She sighed and added wistfully, "but I don't have any plant trips yet."

"Any decent leads at all?" Martin inquired sympathetically.

"Well, maybe one. I got a job application today from a company that seemed interested in me on the phone."

"Good! Send it back quick!"

Myra sighed again. "I would, but there's a problem I don't know how to handle. The application wants a listing of all prior criminal convictions. I want to leave it blank, but I feel guilty."

"Why?"

"Well, I got caught shoplifting once in high school. It was really a stupid thing. My life was still a mess after the divorce, and I was running around with a bad crowd. One of my friends persuaded me to help her lift a sweater from a department store. But we got caught, and we both got a misdemeanor conviction. It was just my luck. The only time I ever stole anything, and I got caught."

"So, what's the big deal? It was high school."

"But I was a senior, 18 years old. I wasn't a juvenile any more. The conviction is on my record forever. Worse yet, we both had a marijuana joint on us. I think I smoked the stuff a total of six times, ever! But we had it right then, so I got hit for possession."

"Ooooh, companies really hate the drug stuff," Martin observed.

"So, I don't see any point in telling them," Myra replied. "Most of them are out of state, and won't check anyway. I did one stupid thing that I'll never do again, because my life was messed up and I hung with the wrong crowd. That's all over now."

"But the companies have a right to honest answers on the application," Martin observed. "They want the right to make the final judgment about who's suitable and who isn't. . . . "

"Martin, you're so picky," Myra broke in. "You know our society, with all those TV shows about cops and criminals. A lot of people are paranoid about crime. You say 'conviction' and they think of a drug-dealing, child-abusing murderer. Some companies have that attitude too. They get wind of even an arrest, and your application is history, no matter how silly the reason. They just don't want to take a chance. And they don't have to in this tight job market."

Martin raised his eyebrows. "The company could fire you if they catch the lie after they hire you."

"I know. Don't get me wrong, Martin. I haven't decided what to do yet. It's a tough problem."

◆ What should Myra do?

CASE 6.3 Using Old Exam Questions

Terence Nonliquet sniffed the crisp night air as he emerged from the theater with his girlfriend Leah Nonlibet. The weather forecasters had predicted the season's first frost for tonight. The brightly twinkling stars in a clear black sky seemed to whisper the same thing. Terence squeezed Leah's hand as he looked up. She smiled and looked heavenward as well. Terence was quick to see the beauty of a vast sky, and she admired that trait in him.

"The sky is gorgeous tonight," she purred. "I'm glad we can share it together."

He remained staring up for another few seconds, then glanced at her and grinned. "And I'm glad we were able to come to some agreement on our time together," he observed. "Regularly scheduled dates don't seem very exciting, but they're like vitamin pills. They keep things healthy when taken without missing." Leah felt her stomach tense a little at the thought of their old argument, but she ignored it.

The two walked slowly to Terence's car. Suddenly a shout pierced the air: "Hey Terence!" Startled, Terence and Leah looked up to see three students rambling down the street in the opposite direction. Terence squinted to see who it was. Then he recognized Arlen, a student in his Comp Sci 109 quiz section. "Hey Terence, looks like you'll be keeping warm tonight!" one of the students snickered as they passed by. The three dissolved into uncontrollable giggling.

Terence rolled his eyes. "Just some of the students I teach in my Friday quiz section," he said to Leah. "They're good guys. Just a little silly tonight, I guess."

"It's OK," Leah offered reassuringly. "It must go with the territory."

"Actually, that class is giving me a headache again. And it's not the students so much as the lecturer."

"Professor Bligh?"

"Yeah. He's giving an exam again next week. You remember how he didn't change the first exam from what he gave years before? Now we'll see what he does this time."

Leah frowned. "Isn't this the guy who lied to you—that he changes the details of the exams from year to year?"

"Yeah. But for the first exam I'm sure he didn't. I overheard some of my students say so, normally weak students who really aced it. They had access to fraternity files or something. Strangely enough, though, those students dropped the class. I don't know why."

"So, maybe there won't be a problem this time, even if the exam matches other years," Leah suggested.

"I don't know. It's possible Bligh might be more careful this time, since I said something to him. But it's also possible that other students have gotten their hands on old files."

"You were going to complain to the department head, Professor Peccavi, last time," responded Leah. "But you never told me what you really did. What happened?"

"Well, I guess not much."

"You mean Professor Peccavi didn't help you?"

"No, I never saw him. I couldn't decide what to do. Finally, I decided to go to him, but by the time I called his secretary for an appointment, he had gone out of town for ten days. She said he was booked up for a few days after that. Now it's been so long that it seems silly to file a complaint."

Leah glanced up with reproach. "Terence, you made a decision by not deciding. Sometimes you think about things way too much."

"Leah, it was a hard problem! Peccavi is a busy guy, and bringing him in would have raised the stakes with Bligh a lot. I didn't want my boss mad at me if I could help it."

Leah shook her head in disgust. "I don't think you did the right thing. Anyway, now you're not as new at the job as you were then. He just blew you off last time when you talked to him, so he's not going to do anything different this time. Terence, I've heard you talk like this before. You know deep down you have to really do something if he's going to change. You have to decide." She paused. Then she murmured, "and it makes me nervous when you can't decide."

Terence looked at her incredulously. "Why? This doesn't involve you!" He paused, and added, "You get nervous even when I do decide! Remember that Zipdraw graphics program I bought using the course account? The one I used in class for a week and then on my own personal reports? You wanted me to clear that with Bligh. When I decided not to, you went into a tizzy."

Leah answered slowly. "Terence, I'm not sure you really decided. You just didn't say anything—maybe out of inertia. As far as I'm concerned, that kind of stuff shows you don't know what you want. Sometimes, especially with important things in life, you have to *know* what you want."

Terence decided to let the subject drop.

◆ What should Terence do?

CASE 6.4 Personal Phone Calls on Company Time

"I was able to visit Monica in the hospital yesterday," remarked Emily Laborvincet to Martin Diesirae as they worked in the computer office at Tripos Metal Polish, Inc.

"Oh? How is she?"

"Still a long way from recovery. She's out of intensive care, and the pneumonia is gone, but the car wreck broke a lot of bones. She'll be in traction for a while."

"It would really help to have her back," Martin remarked wistfully. "This company needs an owner who's actually available. Can't she at least call on the telephone to help out?"

"I don't think so. She looked exhausted. She could only tolerate a ten-minute visit. Her mind is not really with-it yet. Sometimes I didn't understand what she was talking about. Once, out of the clear blue, she said it was time for me to go on a journey. Just like that. When I asked where, she just smiled."

Martin sat silently for a moment. "Sheeze, Emily, look at us! We're not ready to run a company!" he burst out suddenly. He jumped up from his seat and began to pace around. "I mean, I'm just an undergraduate, only a senior. And you! You're a sophomore! We're both only part time, still taking classes. Then suddenly the owner and head of sales get hit by a car, and the company is rudderless." He glared out the door into the rest of the building. "Our technical people don't have a clue how to work with customers, and their eyes glaze over when you talk to them about accounting. Our production managers are the same way. So *we* get stuck."

Emily nodded sympathetically. "But you're a good student. You've won some awards—you told me. You're smart and have lots of energy. And you work hard. You'll be in the work world soon anyway. Why not start now?"

Martin play-acted banging his head against the door. "Because it's too much pressure. I'm interviewing for a job in a tight market. It'll be hard to find one where my girlfriend Myra can get a job too. And she's been a royal pain lately for some reason. And my mother's health has taken a turn for the worse. I have to take care of her a lot of weekends because my dad's out of town on business. He leaves on Fridays, which means that sometimes I have to leave here early those days. And it's a two-hour drive to where they live in Pallortown."

Emily furrowed her brow. "Pallortown? Your parents live there? You don't call there from here, do you?"

Martin straightened up. "Yeah," he said defensively. "Here and there when I need to."

"Martin, I'm in charge of the phone records. It's more than here and there. It probably runs to over an hour a week. Actually, I couldn't figure out what customer we had there. I was going to ask you about it, since you handle customer relations. You know as well as I do that it's against company policy to make long-distance calls on our phones. It costs money, plus there's the lost work time." Martin clenched his fists involuntarily and drew himself up to his full height. Emily now sensed that Martin would not react well to such direct criticism. She had crossed swords with him several weeks before over the arrangement of files on the computer she used for accounting. In the face of an angry outburst, she had backed down. So now Emily broke in to prevent another such outburst. "But I can understand that we might need exceptions, especially in the trying situation right now."

Mollified, Martin relaxed a little. "The situation at home is really bad," he confided. "I worry about it a lot. I work much better when I know things are in order there." He started kicking the door listlessly, face taut with worry.

Emily grew concerned. "Are you OK?"

Martin stopped the kicking and forced a smile. "Yeah, I'm fine. I'm going down to talk with Floyd." He turned and left.

◆ What should Emily do about the phone calls?

7

COMPLETING MORAL JUDGMENTS:
THE DECISIVE ROLE OF INTENTION

"Although it be with truth you speak evil, that is also a crime."

JOHN CHRYSOSTOM (345–407) *HOMILIES*

In the last chapter, we looked at the exterior dimensions of an action in complex situations. Even though we developed a way to balance all the good and bad consequences, we saw that we still needed an extra ingredient to complete ethical judgment. That ingredient was interior intention, which we examine now.

Evaluating Interior Goodness

Each decision we make imprints our tendency to practice the virtues. A good moral decision makes it easier to decide that way the next time, whereas a bad decision makes it more likely that the next decision will be bad. This view focuses on those aspects of interior morality that show up sooner or later in external actions. However, there can be problems with implementing this approach. First, interior effects can be extremely difficult to predict because of the complexity of human behavior. Second, this perspective runs into difficulty when applied to people near death. If our view of ethics remains limited to what can be observed (even in the long run), it doesn't seem to matter very much whether people near death develop their interior character for doing good things in the future! Yet intuitively it seems that interior goodness should still matter somehow. Unfortunately, further progress requires an appeal to philosophical or religious considerations. In line with the approach to ethics we have taken throughout this book, we will not examine those considerations here, as important as they may be.

In discussing interior morality, the systematic framework we developed for looking at exterior consequences offers significant benefits for moral analysis that we don't want to lose. Hence, we will examine attitudes and imprinting on a consequence-by-consequence basis. Let's define our intentions as our chosen attitudes toward exterior consequences—that is, whether we approve or disapprove of each thing that might happen as a result of a moral choice. (Notice that we focus on chosen attitudes here, not on feelings. This distinction is discussed more in Chapter 4.) These chosen attitudes affect our character through imprinting. For example, approving of bad things or disapproving of good ones generally works to weaken our character. Two factors influence the strength of imprinting. Attitudes toward important consequences usually imprint more strongly than attitudes toward minor ones.[1] Also, the degree of approval enters in, with indifferent or conflicted attitudes having effects different from strong approval or disapproval.

Unfortunately, evaluating the likelihood and strength of imprinting in any particular case is quite difficult. As a result, we will not set up a mathematical analogy for imprinting the way we did for exterior goodness with Eq. 6.1.[2]

An Example

To illustrate how to examine attitudes consequence by consequence, let's consider again the example case we discussed at length in Chapter 6. Celia Peccavi had to decide whether to give free breadsticks to her friends, and then whether to lie about what she did to Todd Cuibono. Balancing exterior consequences alone seemed to recommend that she give the breadsticks away and that she lie. Now let's see how to balance the consequences of her interior intentions.

1. Celia's Decision Whether to Give the Breadsticks

With a decision to give the breadsticks, Eq. 6.1 for exterior consequences took the form:

even score	make friends happy	strengthen friendships
(fairness)	(fairness)	(fairness)

Net goodness = (good)(low)(very high) + (good)(low)(very high) + (good)(moderate)(moderate)

Pandarus loses profit	violate rules
(fairness)	(fairness)
+ (bad)(low)(very high)	+ (bad)(high)(very high)

The sum came out to be roughly net zero.

Now, let's list Celia's internal attitudes to each external consequence as shown below. (Of course, we can only guess at what some of these attitudes are!)

External consequences	even score	Make friends happy	Pandarus loses profit	stronger friendships	violate rules
Celia's approval	approve	approve strategy	disapprove weakly	approve strongly	disapprove

Now, we need to guess how Celia's attitudes will strengthen her continued practice of the virtues (if she proceeds with the action she's thinking about). We can list interior consequences for imprinting just below Celia's attitudes, as shown below. If an imprinting entry does not refer directly to one of the virtues, it may be helpful to add in parentheses the virtues most involved. If the virtue is justice, it can help to list which part of justice (truth or fairness) is affected.

External consequences	even score	make friends happy	Pandarus loses profit	stronger friendships	violate rules
Celia's approval	approve	approve strongly	disapprove weakly tolerate	approve strongly promote long-term	disapprove tolerate
Internal consequences	promote social cohesion (fairness)	promote social cohesion (fairness)	manifest unfairness	social cohesion (fairness)	manifest unfairness

An item here is the restaurant's loss of money. The amount is only a few dollars, so the exterior importance remains low. However, if Celia decides to give the breadsticks away (as she actually did in the story), her willingness to accept such obvious unfairness to the restaurant would be a serious moral concern. Her violation of reasonable, clearly presented rules is also a major problem.

With a decision to withhold the breadsticks, Eq. 6.1 became:

$$\underset{\text{(fairness)}}{\text{disappoint friends}} \qquad \underset{\text{(fairness)}}{\text{weaken friendships}} \qquad \underset{\text{(fairness)}}{\text{lose friendships}}$$

exterior goodness = (bad)(low)(very high) + (bad)(low)(moderate) + (bad)(moderate)(low)

The sum was somewhat negative.

The interior analysis takes the following form:

External consequences	disappoint friends	weaken friendships	lose friendships
Celia's approval	disapprove	disapprove	disapprove

Internal	inhibit	inhibit	inhibit
consequences	social cohesion	social cohesion	long-term social
	(fairness)	(fairness)	cohesion (fairness)

In contrast to the option of giving the breadsticks, here there seem to be only mild negative interior consequences. Comparing the two options from the interior perspective suggests that giving the breadsticks is worse than withholding them. This result is the opposite of what we concluded on the basis of exterior consequences.

2. Celia's Decision Whether to Tell Todd the Truth

We saw that with a decision to tell the truth, Eq. 6.1 could be cast in the following way:

| Todd tells boss | boss causes trouble | Celia fired | Celia broke |
| (?) | (fairness) | (fairness) | (fairness) |

Net goodness = (?)(zero)(very high) + (bad)(moderate)(high) + (bad)(high)(low) + (bad)(high)(low)

example to employees
(truth, fairness)

+ (good)(moderate)(low)

The sum yielded a modest negative. From an interior perspective, we can make the following educated guesses about Celia's intentions:

External *consequences*	Todd tells boss	boss causes trouble	Celia fired	Celia broke	example to employees
Celia's *approval*	disapprove	disapprove	disapprove	disapprove	approve weakly
Internal *consequences*	inhibit truth	inhibit fairness	inhibit fairness	none	promote fairness

Notice how the exterior consequence of Todd's telling his boss, which had no real moral importance in an external sense, takes on moral importance in an internal sense. On balance, Celia's intentions are more bad than good.
  With a decision to lie, Eq. 6.1 took the following form:

| Todd deceived | Todd suspects | Celia discovered |
| (truth) | (truth) | (?) |

Net goodness = (?)(zero)(moderate) + (bad)(very low)(high) + (?)(zero)(very low)

several mostly bad consequences
of very low probability

+ (bad)(low to high)(very low)

The sum yielded a low net negative. An interior analysis yields the following result:

External consequences	Todd deceived	Todd suspects	Celia discovered	several mostly bad consequences of very low probability
Celia's approval	approve	disapprove	disapprove strongly	disapprove
Internal consequences	promote manifest untruth	promote untruth	promote untruth	promote unfairness

Again, attitudes toward exterior consequences of no moral importance can take on significant interior importance. Here we see that if Celia decides to lie (as she actually did in the story), her attitudes would promote a clear-cut willingness to deceive and would be a serious moral negative.

In summary, lying is a significantly stronger interior negative for Celia than telling the truth. Once again, the conclusion is reversed from the result of the exterior analysis.

Balancing Evaluations of Interior and Exterior Goodness

Now that we have seen how to balance consequences separately in the exterior and interior realms, how can we put these evaluations together into a single coherent judgment? Unfortunately, there seems to be no single, straightforward criterion to use. The question is a complicated one that ethicists still debate vigorously.

However, one possible ingredient is the relative number of people affected by the exterior consequences and the interior ones. Often the number remains quite small for exterior consequences. However, interior imprinting can show up in all our interactions with others—potentially leading to far larger effects. Such effects become particularly large for public servants. On the other hand, numbers do not tell the whole story. The degree of impact also plays an important role. A fourth-grade teacher comes into contact with far fewer people than does a grocery checker, but the teacher has far greater influence on moral behavior.

To apply this criterion in the case of Celia, we must try to guess at her effect on others. Celia does not live alone like a hermit, but she doesn't hold a position of public trust either. Thus, the imprinting effects of what she does are about average, meaning significant. In view of the very poor interior consequence of offering the breadsticks, withholding the breadsticks becomes the best option. For similar reasons, telling Todd the truth also becomes the best option.

The "Solomon Problem"

What happens when, after adding in intention, our analysis yields two "best" options? In honor of the wise king of antiquity, we might call this difficulty the "Solomon problem."[3] In addressing the Solomon problem, we must draw upon a fundamental principle whose origins go back to the ancient Greeks. The basic idea may seem intuitively reasonable, but requires appeals to philosophy or religion for a more complete defense.

Principle: *The obligation to avoid what is bad outweighs the obligation to do what is good.*

We find evidence for this principle in the long-held rule for the physician confronting a sick patient: "Do no harm." Moreover, early systems of law like the Code of Hammurabi and the Ten Commandments spoke more often about which evils to avoid rather than which goods to pursue. This tradition continues to the present day; most codes of law tend to say much more about what people should *not* do than about what they *should* do.

When applied to situations where two moral options seem on balance to be equally good (or bad), we should try to pick the one with the fewest really bad consequences, even if compensated by consequences that are extremely good.

Cooperating in the Evil of Others

Moral questions have special urgency when we find ourselves pressured to cooperate with an evil pursued by someone else. For example, suppose you live next door to a homeowner who has been charged several times with stealing money from his employer, but has always escaped conviction on some technicality. Little question exists in the minds of yourself and other local residents that he is a criminal. One day, he asks you to mow his lawn. You need the job, so you agree. Now suppose this neighbor asks you to mail a letter after you finish mowing—a letter you suspect contains fraudulent forms. You mail the letter grudgingly. Suppose that the next time this happens, you decide to confront your neighbor. He admits what you suspect but threatens to stop hiring you unless you mail the letter. You resentfully give in and mail the letter. Suppose finally that with the next letter your neighbor takes back his threat and instead offers you a large bribe. You mail the letter and accept the money. At what point has your behavior crossed the line into wrongdoing?

The perceptions of others must be taken into consideration also. There can be situations where someone can perform an action that by itself would not be wrong, but where others might misunderstand what is happening and conclude that the action is wrong. For example, mowing the lawn of your criminal neighbor might not be wrong, but others might think that you are trying to profit from his criminal activity. Because this is an unreasonable judgment to draw from your action, you should probably still go ahead and mow the lawn. But if others were to see you mailing your neighbor's letter (even if its contents remained unknown to you), they could more reasonably conclude that you are helping his criminal activity, and you should probably refuse to do it.

Such situations come up so often and are so hard to resolve that classical moral theory has developed a special vocabulary for classifying degrees of cooperation. This classification introduces no new principles from what we have already seen, but provides a quick and convenient means for looking at such problems. The classical view assigns three categories:

Mediate material cooperation: Here we disapprove of the injustice we see another doing. Furthermore, our own actions would normally be considered good or neutral. Also, our actions should provide nothing necessary for that injustice to occur, and should be only remotely involved with the evil taking place. Mowing the lawn for your embezzling neighbor would normally fall into this category. Classical moral writers generally consider this kind of cooperation to be acceptable given adequate reasons.

Immediate material cooperation: Here also we disapprove of the injustice we see another doing, and under ordinary circumstances our own actions would again be considered morally good or neutral. However, now our actions provide something needed for the evil to occur. Reluctantly mailing the fraudulent letter because of threats from your neighbor falls into this category. Classical moral writers suggest that such cooperation is acceptable only for serious reasons. Recourse to our more general analysis is usually required to determine whether the reasons are sufficiently serious. Crucial factors include the precise importance of your role in the injustice, the degree of harm caused by not cooperating, and the possibility of misleading others about what is happening or giving a bad example.

Formal cooperation: Here, we now both approve of the evil and provide through our actions something needed for it to occur. Furthermore, under ordinary circumstances our actions would be considered morally bad. Mailing the fraudulent letter in exchange for a bribe falls in this category. Classical moral writers consider this kind of cooperation to be unacceptable.

A REAL-LIFE CASE: The Problem of Performance Evaluation—Grade Inflation

Evaluation of employee performance has long occupied a central place in corporate life. Scientists and engineers routinely (usually annually) undergo performance reviews, and often find themselves compared directly with their peers. Many divisions, departments, and even entire companies rank their employees on a single scale of best to worst. The assignment of such rankings raises many questions on both exterior and interior levels. Of course, college students have long experience with these issues—through the grading system. Interestingly, much evidence has accumulated to show that grade inflation has become common across the United States during the past few decades. For example, at one well-known university, 69 percent of grades fell between B− and A+ during the period 1973–1977. That proportion increased to 83 percent during the period 1992–1997.

Several motives may drive the increasing leniency of professors in their grade assignment. In an era when student course evaluations make up an increasingly important part of promotion and tenure decisions, younger faculty may avoid grading strictly. The entry-level job market has become increasingly competitive, and faculty may not want to put their students at a disadvantage. Student financial aid packages and insurance discounts are often tied to grade point average; professors may feel reluctant to deprive students of these benefits. Finally, many busy faculty quickly learn that easy grading translates directly into fewer complaining students at the office door.

On the other hand, regardless of the intentions, grade inflation has negative exterior consequences. If everyone gets A's, it becomes difficult to tell which students are really the best. After a few bad experiences, however, many employers and graduate schools learn which universities have grade inflation, and become suspicious of hiring or admitting any of their graduates. Finally, grade inflation sends the hidden message that performance does not really matter. Students who internalize this message face a rude shock in the work world, where performance is king.

- ◆ Do you think grade inflation is a serious problem?
- ◆ What steps can be taken to reduce the negative effects of grade inflation?

♦ To what extent might students be responsible for grade inflation?

♦ Both good and selfish intentions may underlie faculty tendencies to inflate their grades. On average, which motivations do you think are strongest or most common?

References

Frerking, Tim. "Now My B's Are No Better Than My C's? When a B Isn't a B." *Iowa State Daily*, 24 June 1997.

Tello, Angie. "Grade Inflation Sweeps Colleges Nationally." *The Lariat* (Baylor University), 5 March 1998.

"To do evil that good may come of it is for bunglers in politics as well as morals."

WILLIAM PENN (1644–1718), *SOME FRUITS OF SOLITUDE*

Notes

1. Exceptions do occur, however. Sometimes small events influence the interior life profoundly. Such changes are rare, but they do happen—a small deed opens our eyes to a whole pattern of behavior we never really thought about before.

2. As mentioned in note 2 of Chapter 6, the method we outline here bears certain superficial resemblances to utilitarianism. The casting of interior morality in the form of consequences is an example. However, standard utilitarianism largely ignores interior morality and character virtue.

3. This terminology alludes to a legend about the biblical King Solomon. Solomon was so renowned for his wisdom that the Queen of Sheba decided to pay him a visit to test his abilities. During their meeting, she presented him with two bouquets of flowers. One was real, but the other was an exquisite artificial copy made by expert craftsmen, who even added perfumes to reproduce the floral scent. The queen offered Solomon extravagant gifts if he could distinguish the real flowers from the fake (i.e., properly choose between two apparently identical options) without touching the bouquets. Even after lengthy inspection and considerable pondering, the wise king despaired of choosing the right one. He was about to concede failure when he noticed a honeybee attempting to enter the room despite attempts by the court attendants to wave it off. Solomon's eyes lighted up, and he motioned the attendants to let the bee enter. The bee buzzed directly to the bouquets, alighting first on the artificial one. After a moment's hesitation the bee took off disconcertedly and flew to the real one. Immediately it buried itself in the flowers, busily collecting nectar as bees usually do. King Solomon smiled and pointed to the bouquet that had so attracted the lowly honeybee. The queen, now convinced of the king's wisdom, approved the choice and gladly gave him the gifts she promised.

Problems

1. Write a paragraph or two describing an ethical dilemma you have encountered in a job you've had. (If you've been lucky enough never to have been confronted with a problem like this, describe one that a friend or relative of yours has had.) Analyze the case as follows:

    a. List the options/suboptions available to the person who had to make a decision, together with the event tree flowing from each option. Indicate your estimate of the probability for each consequence on the tree (high, moderate, low).

    b. Write down an expression for exterior goodness according to Eq. 6.1 as shown in the text example.

    c. List for each consequence the person's intention and likely imprinting effect as in the text example.

    d. Recommend what you think the person should have done. Note: you don't have to say what was actually done in real life (unless you want to)!

2. Each case below has a question after it.

    a. List the options/suboptions available to the main character who has to make a decision, together with the event tree flowing from each option. Indicate your estimate of the probability for each consequence on the tree (high, moderate, low).

    b. Write down an expression for exterior goodness according to Eq. 6.1 as shown in the text example.

    c. List for each consequence the main character's intention and likely imprinting effect as in the text example.

    d. Recommend what you think the character should do.

CASE 7.1 Reading What Is on Others' Desks

"Wow, your desk is a mess!" exclaimed Leah Nonlibet to her boyfriend Terence Nonliquet. The two were in Terence's office in the computer science department at Nosce te Ipsum University, where Terence served as a teaching assistant for a quiz section. It was the day after Thanksgiving, and Terence was desperately trying to catch up on some work. Leah had accompanied him for togetherness' sake.

"Yeah, I haven't had a chance to sort through all that junk," Terence sighed, glancing up from his computer terminal on a table adjoining the desk. He stared wearily at the huge pile of papers that covered every square inch of the desk top. "A lot of it is important, too—unopened mail I've been getting here at work since last week.

A lot of it is software ads. I also brought in the mail I get at home, thinking that I could just open everything at once." He paused. "But it didn't happen that way. Say, Leah, maybe you could help. Can you try to separate the mail from the other junk? Maybe you can put the stuff that looks most urgent on top."

Glad for something to do, Leah got started eagerly. "I bet your office hours weren't too busy this week," she ventured for conversation.

"No," he replied, face buried in the computer screen in front of him. "I think a lot of students just took the whole week off. Professor Bligh said his lectures were only half full."

Leah nodded, and then frowned as she came across a letter that had apparently arrived at Terence's office on Wednesday. It bore a woman's handwriting, and listed "Celia Peccavi" on the return address. "The department head for computer science is Professor Peccavi, isn't it?" she asked.

Terence did not look up. "Yeah. He's a pretty nice guy, but busy. I thought I told you once, his niece is in my quiz section. Her name is Celia. . . ." Terence engrossed himself again in the task in front of him.

Leah sniffed the air as she held the envelope, and suddenly realized that it was scented. She frowned again, glanced at Terence to ensure he was busy, and held the envelope up to the light. The paper was thin, so she could read the short letter inside with little trouble.

It read, "Dear Terence, I found out just before I left for Thanksgiving vacation that my dorm floor at Boogie Hall is having a party Sunday night when I get back, at 7 p.m. We're having games, with the same teams that played in the volleyball tournament six weeks ago. Since you were nice enough to be on my team, I wanted to invite you to the party. I don't have your home address or phone number, so I'm writing you at work. If you come, just ask the hall office to ring my room and I'll come down and get you. Hope you can make it! Celia."

Leah's stomach twisted with fear and anger. Her eyes riveted on the little heart Celia had used to dot the 'i' in her name. She put the letter down and sat tensely for a minute. "So, Terence, how well do you know this Celia?" she inquired warily.

A long pause ensued. "Oh, not very," Terence finally replied, not looking away from the computer.

Leah clenched her fists involuntarily, then relaxed them. She again glanced at Terence to ensure he wasn't looking, and slipped the letter into a pile of advertising on the desk. Then she quickly piled the rest of the mail into a neat stack. "OK, Terence, I've stacked your mail," she announced. There's only one or two things I think you need to see. They look like bills. I left them on top."

"Thanks," Terence mumbled.

Leah smiled to herself. "I'm going to the bathroom," she said, not completely able to hide the bitterness in her voice.

"OK," replied Terence, oblivious to her tone.

◆ Should Leah continue to hide the letter from Terence?

Case 7.2 Dating the Boss's Relative

As he sat cleaning the desk in his office, Terence Nonliquet heard a tentative knock on the door. "Sheeze, the semester is over," he said to himself. "I thought I was finished with this TA business until next semester. I hope no one wants to complain about their grade." He straightened up in his chair. "Come in," he called. Celia Peccavi, a student in his quiz section, timidly opened the door. "Oh, it's you Celia," he muttered with barely concealed surprise. Then he recovered and his voice strengthened. "Come in. I didn't expect to see you here. Your grade was pretty good—you're not here to complain, are you?"

"Well, no, I just wanted to see what I got on my final," she replied softly, brushing her long hair away from her face with a deft stroke.

"Pretty good, but I'll have to check the exact number. Just a minute," Terence responded, fumbling through some papers. "Here it is . . . 81 percent. That got you a solid B." He held the paper up. "I can show it to you if you want," he offered.

Celia shook her head. "No thanks." A pregnant pause ensued. "My dorm floor had a killer party a couple of weeks ago," she ventured. "Too bad you couldn't make it."

Terence shifted uneasily in his chair. "Oh, that . . . oh . . . Celia, I was meaning to apologize," he said guardedly. "I didn't see your letter until yesterday. It got lost in this pile of junk," he continued rapidly, waving his hand at the papers on his desk.

Instantly Celia straightened up. "So you didn't get it until yesterday?" she said with a faint smile.

"No, I'm really sorry. It just got lost. I don't know how it happened. It got separated from my other mail somehow."

Celia unzipped her jacket and moved closer to lean her hips on his desk. "So you would have come?"

Terence retreated slightly in his chair and averted his eyes. "Well, I . . . uh . . . I don't know," he stammered. "I don't remember what my schedule was."

Celia saw her opportunity, and seized the initiative. With sultry deliberation, she shed her coat, hopped up on his desk, crossed her legs, and smiled down on Terence. He tried to maintain a professional manner, but his eyes remained riveted on her every move.

"Terence," she began soothingly. "I can understand how you feel. You've been nervous because I was a student in your class. You had to be careful. I respect you for that."

Terence relaxed slightly. "Yeah," he rejoined uneasily.

"But that's over now, Terence. The class is over. So you don't have to worry about appearances if we do something together."

Terence bolted up straight. "Do something? What do you mean, 'do something'? I have a girlfriend."

A slight chuckle escaped Celia's lips as she recrossed her legs and adjusted the hem of her skirt just above her knee. "Oh, yeah. I forgot. Leah, right? You talked about her in class. I've seen you walking with her." She wrinkled her face. "She looks and sounds like a worrywart."

"She's very nice," Terence countered, somewhat hollowly, his eyes returning to her knee.

"I'm sure she is. You wouldn't date just anyone, would you?" Celia purred. "Anyway, she doesn't have to know, at least for a while."

Terence drew a deep breath and summoned his remaining defense. "Celia, I don't think this is right. I'm still a TA next semester. For Comp Sci 110. It comes next in the sequence after 109—what you took this semester. Your uncle is the department head. He's in effect my boss. It's just too weird."

"My uncle? He won't mind. He thinks you're a great guy. He told me so. I know he'd rather see me go out with you than lots of other guys I've dated." Celia paused for effect. "And I'm his only niece. He thinks the sun rises and sets on me. He knows how to take care of people who are nice to me."

"Celia . . ." Terence protested weakly. "You're nice, but . . ." His voice trailed off.

Celia's voice sharpened as she played her final card. "Terence, if it's not obvious, I like you. Lots of guys would give their right arm to go out with me—but not you. You're too worried about 'honor' and all that stuff. But I don't give up easily. You won't get rid of me just because the semester is over. I want you to go out with me— just once over Christmas break. Once is enough, because I'll see to it that you want more. But if not, I promise—just once. No more." Celia leaned over toward Terence to drive her point home. "If you don't give me just one date, I'll enroll in your class next semester. I have room in my schedule, and the section is not full. I checked. Then you'll have to look at me for another whole semester." She straightened up with a haughty grin. "But you'd like that, wouldn't you?," she added, glancing down at herself.

"Celia, that's like . . . blackmail," Terence murmured.

"No it's not. No one's forcing you to do anything. I can take any class I want. If you don't give me one date, I'll just want *your* class."

She stood up and turned to go, tossing her hair with a flourish. "Don't worry, I'll be a good girl. You won't have to keep me after school!" she laughed.

◆ What should Terence do?

CASE 7.3 Dating among Coworkers

Emily Laborvincet and Martin Diesirae sat sipping coffee in the manager's office at ripos Metal Polish, Inc. It had been a long day, but the two of them had averted several serious crises. They felt tired but satisfied.

"So, Martin, enough business," said Emily. "How are things going at home? You know—with your mother. Is she getting better?"

"A little," Martin sighed. "And my dad has cut down on his traveling a little, so I haven't had to go home to care for her as much. Still, the whole situation takes a lot of my energy."

"I figured as much," responded Emily sympathetically. "I noticed today when doublechecking our accounts that the company phone bills still show a lot of calls to Pallortown."

Martin smiled sheepishly. "I told you, I can work better when I can call when I need to. Thanks for letting it slide."

Emily smiled back fleetingly, but then grew serious again and changed the subject. "How did your plant trip go?" she inquired. "You were gone early this week and never told me about it."

"You mean to Motorel? It was pretty good. The facilities are impressive. It seems like a good group of people, and the job they're trying to fill looks interesting."

"Do you have any other trips lined up?"

"Not so far. But I've got so much going on with classes and this job that I haven't been able to concentrate on the job search as much as I want. Maybe spring will be better."

"So would you take the Motorel job if they offered it?"

Martin frowned. "I don't know. There would be a lot of things to work out. A lot depends on how much I work here next semester. If I work full time to help bail the company out of the mess it's in, I'll have to delay my graduation till next December."

"And your working will depend on whether Monica and Chase can recover from the car accident, so you won't have to fill in for them?"

"Partially, but that's not the only thing. This company means a lot to me, but there are other things to worry about, too. My girlfriend Myra has gotten just impossible, lately."

Emily perked up. "Oh?"

"Yeah," Martin continued wearily. "She's a high-maintenance

woman. Always depressed or crying about something. All her complaining gets me down."

"How does that affect the job with Motorel?"

"Myra is also supposed to graduate this spring, in environmental engineering. She's not sure she can find a decent job in East Cupcake, Idaho. Motorel is really the only big employer there."

"The old two-body problem, huh?" Emily offered.

"I guess. But it's more than that. Motorel is just a symptom. The problem is deeper. Myra's had a hard life. Her parents split up in a nasty divorce. She had some minor brushes with the law."

Emily raised her eyebrows. "The law?"

"Yeah, minor stuff. But they're convictions on her permanent record. She lies about them on job applications. It could be a time bomb someday. But that's not the worst of it. She's sad a lot, and always seems to be dragging around. She has a hard time trusting men because her father let her down a lot. And she lets people push her around all the time. She's like a doormat. Sometimes it's hard to respect her."

Emily smiled. "So you want someone with more spine, huh?"

Martin looked up at her. "Yeah. Someone with energy and brains, too."

Still smiling, Emily twirled her hair and glanced out the window. Then she became more serious. "I'm having trouble with my boyfriend, too," she remarked.

"You mean Todd?"

"Uh-huh. He has good traits, but he's very self-centered. Always wants to know what's 'in it for him.' I'm not like that. I mean, you have to look out for yourself, but that's not all you can think about, right?"

"Right!," Martin said earnestly. "Are you going to break up with him?"

"I don't know. It's nice to be with someone. You know, to talk and stuff." Her glance returned to fix itself on Martin. A long pause followed.

Martin broke the silence. "Like I said before, I'm starved. You want some dinner? I know a new restaurant that just opened a week ago. It's just a few miles from here."

Emily smiled broadly, then crossed her arms in mock seriousness. "Martin Diesirae! Is this a date?"

"I don't know," he shrugged impishly. "Call it what you want. You said you wanted someone to talk to. I don't mind talking." Martin drew closer to her. "Especially to someone with nice eyes."

Emily blushed and looked down. But then she became serious. "Martin, this is dangerous. I mean, neither of us has broken up yet. And what if anyone at Tripos finds out? We'll never hear the end of it! And what if it doesn't work out? We still have to do a job together. That could be uncomfortable."

Martin sat back in his chair. "Yeah," he agreed. "But I just asked for dinner. Is that so bad?"

◆ Should Martin and Emily pursue this relationship?

CASE 7.4 Publicly Misrepresenting Private Attitudes

"These long lines are just impossible!" grumbled Terence Nonliquet to his girlfriend Leah Nonlibet as they waited to pay for their textbooks at the Nosce te Ipsum University Bookstore.

"Yeah, it's always bad right before the semester starts," Leah agreed.

Terence glanced at his watch. "I've still got a lot of work to do today! Professor Bligh is having a TA meeting tomorrow, to go over some last-minute details before classes start."

"Too bad you have to deal with that guy again this semester," Leah ventured sympathetically. "He gave you enough trouble last fall."

Terence tensed at the memory. "Uh-huh. And I know I'm not on his good side. It's bad enough that I'm only an undergraduate, and a junior at that. But he *really* didn't like when I gave out those copies of exams from previous years."

Leah became exasperated. "Terence, you did about the worst thing in that case. You knew he was using the same exams from year to year, and that some people had access to them. When you found that out after the first exam and he lied to you about doing it, you could have gone to the department head. But you waited around until it was too late. Then on the second exam, you delayed until the last second. Then you disobeyed Bligh's orders and handed out an old exam to whomever came to your office hours. But not everyone came to your office hours, so it was still unfair to some students. And it was a dumb idea to tell your students to keep the whole thing quiet. No big surprise—Bligh found out from some bigmouth anyway, and got mad. And you still never told the department head! So the exam was even more unfair, you got Bligh mad, and you had no one to back you up!" Leah shook her head. "I hope you don't pull that one again!" she scolded.

"Leah, let's not talk about it," Terence responded testily. "Anyway, I've got new problems to worry about with Bligh."

Leah rolled her eyes. "What this time?"

"We had our first TA meeting yesterday. He started telling us about this software package he wants to try next semester. It's called Witch. He says it's great for a class like Comp Sci 110, and that the students will love it. It does all kinds of calculations, equation-solving, and graphics. He's having it installed on the department's computers. But it's affordable enough so that the students can buy it for their own computers if they want."

"So, what's wrong with that?"

"Well, I happen to know there's another package like it that another company just put on the market. It's called Wizard. I tried it at a friend's house the other day. It's about the same price as Witch, but it's a little faster and more powerful. And I think the commands are easier for students to understand. So I brought this up at the meeting, and suggested we use Wizard instead of Witch."

"And?"

"Well, first Bligh tried to pooh-pooh Wizard. He said it wasn't all it was cracked up to be. I said that might be true, but on the basis of my experience, I knew it was better than Witch. He said that as an undergraduate, I was too inexperienced to make such a judgment. But then one of the graduate student TA's—Veronica—jumped in and said the same thing I did. And it turns out Bligh hasn't ever used Wizard, so he doesn't really know what it can do. So he started to get mad."

"What did he do?"

"He told us that we were just going to use Witch for the course and that was that. And he said we should recommend Witch to students who had their own computers."

"You mean, instead of Wizard?"

"Yeah. And that's where things got ugly. It's one thing to pick a piece of software for a course. It's another thing to tell TA's to advertise it—in preference to products that are better! Veronica and I started criticizing Witch again. Then somehow in the conversation it came out that Bligh's brother is high up in the company that makes Witch. Worse yet, one of the TA's let out that Bligh had told her he owned stock in the company. That made it sound like Bligh is making us do this for his personal gain—and for his brother."

Leah recoiled in shock. "No!" she exclaimed.

"Yes! Then Bligh got madder than I've ever seen him. He started swearing, and yelling that he was the professor and we were only students, that we didn't know what we were talking about, and how dare we suggest that he had other motives. And he gave us a direct order to recommend that our students buy Witch for their computers."

"So what did the TA's say?"

"Actually, except for Veronica and me, I think most of them don't care all that much, so they'll do what he says. I talked to Veronica later, and she said Bligh is on her doctoral committee. So she doesn't want trouble. I guess Bligh will get his way, except maybe for me."

Leah's eyes widened. "You're not going to tell your students something you don't believe, are you?"

Terence shrugged. "I don't want to. But I have enough trouble with Bligh as it is. Witch will be used for the course, so the students should buy the same program if they want to use it for homework."

"But what if one of them asks you about Wizard? They might, you know!"

"I don't know."

"Terence," Leah sighed, "I hate to say it again, but you always say that when there's a hard decision to make."

◆ What should Terence say to his students?

8

MORAL RESPONSIBILITY

"Freedom! A fine word when rightly understood. What freedom would you
have? What is the freedom of the most free? To act rightly!"

JOHANN WOLFGANG VON GOETHE (1749–1832), *EGMONT*

Factors Limiting Moral Responsibility

In classical moral thought, morality concerns the goodness of voluntary
human conduct that impacts the self or other living beings. The word "voluntary" holds great importance, implying that we must have adequate control over what we're doing. *Assuming we have not deliberately allowed
ourselves to remain ignorant, powerless, or indifferent, we have complete
moral responsibility for what we do with adequate knowledge, freedom, and
approval.* This definition points to three factors that can limit moral responsibility. Let's examine each of them.

Lack of Knowledge

Lack of knowledge that limits moral responsibility takes two forms. Through
no fault of our own, we might be ignorant of certain key aspects of a situation. For example, a man may pay for groceries with counterfeit cur-
rency, not knowing the money is fake. We may also be unaware that a
certain kind of action is unethical. For example, a very young girl may
take her mother's lipstick and draw all over the household walls with it.
Both kinds of ignorance can arise from several sources. Sometimes, ignorance simply happens without fault on anyone's part. We may not get the
opportunity to learn what we need to know, or we may simply forget. In
other cases, however, ignorance comes about from the negligent or fraudulent behavior of others.

Lack of Freedom

Lack of freedom often occurs due to factors external to the self. For example, a man might be driving a car with faulty brakes and accidentally

hit an elderly lady as a result. Threats of physical or psychological violence by others can remove freedom in a similar way.

Lack of freedom can occur due to internal factors as well. Actions taken under the influence of alcohol or drugs fall into this category. Deeply ingrained habits can also remove freedom. To fully do so, however, the habits must normally be at the level of physical or psychological addiction. Finally, overwhelming passion can remove moral responsibility. Sometimes in very trying circumstances, we become so overcome with feelings of fear, guilt, or grief that we lose control over what we do.

Lack of Approval

Many moralists call a lack of complete approval "deficient consent." However, "approval" carries an active meaning missing from "consent," so we will use "approval." Regardless of the terms we pick, however, the basic idea is that we have responsibility only for what we do willingly. An inability or unwillingness to give approval to what we do can happen in several ways, but most commonly through insufficient time for reflection. Some moral situations, particularly complex ones, require lots of time to think through. If circumstances force us into a snap judgment, we do not have time to think through all the consequences, and cannot really approve or disapprove of them.

Notice in these paragraphs that we have restricted responsibility to situations in which we have knowledge, freedom, and approval. However, earlier we carefully qualified our limitation of the word "voluntary" with the phrase "assuming we have not deliberately allowed ourselves to remain ignorant, powerless, or indifferent." If we have deliberately or carelessly compromised our knowledge, freedom, or approval, we remain responsible for what happens. Thus, in many legal liability cases, a key question often revolves around what a "reasonable person" exercising "due diligence" should have known or done. Of course, such assessments can be difficult to make. Issues of freedom become complicated. Sometimes people lose control over their actions because of psychopathologies, addictions, and the like. The unconscious mind plays a key role in such cases. However, one can use the conscious mind to do things that help to shape the unconscious mind and limit its harmful effects. Obvious examples include participating in psychotherapy or addiction recovery programs. As a result, the degree of guilt of the psychologically unbalanced, the drug addicted, and the abused for their actions remains a matter of vigorous debate.

Degrees of Responsibility

Knowledge, freedom, and approval can all exist in varying degrees. Thus, moral responsibility falls along a continuous line, ranging from none to partial to complete. The legal system in the United States reflects this idea in laws governing civil liability. While details vary from state to state, the legal concept of "comparative negligence" concerns partial responsibility.[1] When an insured person has an accident and files a claim with the insurance company, disputes over whether the company must pay sometimes require settlement in court. The court must decide whether negligence of the claimant contributed to the accident, and if so, must assign a number to how much. In states whose laws specify payment according to "pure comparative negligence," the payment required of the insurance company decreases as the negligence of the claimant increases. For example, if the court decides that claimant negligence contributed 80 percent to the cause of the accident, the insurance company needs to pay only 20 percent of the claim.

It should come as no surprise that such a numerical analysis can prove extremely difficult. Some states specify that the court must determine a likely *range* of negligence for a given accident. For example, a court might find a claimant's negligence contributed between 20 and 30 percent to the accident's cause. In states with some versions of "modified comparative negligence," if the claimant's range exceeds 50 percent in its entirety (say, 55% to 65%), the claimant gets nothing. The idea is that if a person has more than about half the responsibility for an accident he or she deserves no payment. Obviously, quantitative approaches to moral responsibility vary considerably. However, in civil law, where responsibility for accidents, environmental cleanup, and defects in medical products translate qualitative liability into quantitative payments, the principles we invoke for making such assessments become very important.

An Example

Let's see how some of these ideas play out in an actual situation.

CASE 8.0 Responsibility: Invincible Ignorance

"But I didn't mean to burn the filament out!" sputtered Leah Nonlibet.

"You didn't know what you were doing, and now it's busted," growled Bernard Orson, the graduate student in charge of the mass spectrometer. "I don't have any spare filaments in stock, so it'll take me a week to order a new one. Then it'll be a day to actually take the instrument apart and replace it!"

Leah became irritated. "It's not my fault you don't have any replacements. It seems a spare should always be lying around, because you never know when a filament will burn out. Anyway, I was only trying to help." This was not the first time Leah had argued with

Bernard. Bernard had just begun his second year as a Master's degree student working for Professor Clark. He had shown Leah some basics of how the mass spectrometer worked so she could perform routine maintenance tasks like changing the oil in the pumps.

"I thought you were paid to keep stuff like this in stock," retorted Bernard.

"Look, Bernard, I'm only supposed to be here eight hours per week. Yes, part of my job is ordering new supplies, but I can't order every single thing that every person needs. That would take more time than Professor Clark is paying me for. Plus, he assigns me a few of my own experiments. I order things that people ask me to, and then if I have time I try to look around to see what else is needed."

Bernard continued to glare at her. "I told you not to touch the mass spectrometer controls when I'm not around. I had the ion source bias current turned down artificially low because I was trying to look for an unstable parent species. That's why the emission current looked so low. At least you could have looked to see whether we had some spare filaments lying around before you took your chances with this one."

"Bernard, I told you, I was trying to help. I knew you were doing some signal averaging for an important experiment. When I happened to walk by and noticed the emission current was low, I tried to look for you. But you and everyone else were out to lunch. I thought that maybe the emission current had just drifted like it sometimes does, and that you'd be happy that I fixed it and saved you from a wasted experiment. That's when I slowly turned up the current to the filament to crank up the emission. I really didn't turn it that far before the filament blew. You told me a few weeks ago that the filament was old and living on borrowed time anyway. It was probably too fragile to handle any extra stress."

"Don't touch my instrument when I'm not around!"

"It's not your instrument, it's the lab's. Anyway, I don't think you should leave an experiment like this unattended."

"You're not paid to tell me how to do my experiment!" Bernard shot back. "I'm going to tell Clark that you blundered and broke my equipment, and that in compensation I want him to assign you to plot up the data graphs for my talk at the American Geophysical Union conference in a couple of weeks."

"What, you want me to be your personal secretary? I took this job to get some lab experience, not to play around in an office with spreadsheets! Anyway, I have obligations to other people in the lab, too. What makes you think Professor Clark will do what you ask?"

"He's pretty easygoing, and wants to keep people happy. If you tell him you agree to do the presentation work, he'll say 'fine.' Presentations are part of research, too. Anyway, you owe me. If you don't do this for me, I'll see to it that Clark hears about every mistake and omission you make. Then he'll think twice about what to write when you apply for graduate school."

♦ Was Leah responsible for breaking the filament? To what degree?

♦ Should Leah agree to do the spreadsheets?

We can list the factors influencing Leah's responsibility for the accident as follows:

Factors increasing responsibility	*Factors decreasing responsibility*
Knew that she was inexperienced	Filament was old and fragile
Did not check for spare filaments	Equipment only appeared to need attention
Had not ordered spare filaments	Did not intend to break equipment

Let's apply our threefold test of knowledge, freedom, and approval. It's clear that Leah didn't know that the apparatus was actually operating the way Bernard had intended. Leah also lost freedom to some extent because the filament was old and fragile, thereby constraining her ability to tweak the instrument's knobs to fix things. Finally, it's clear that Leah did not approve of the filament's breakage. Thus, from this limited perspective Leah lacks not just one but all three of the ingredients for full moral responsibility.

However, from a broader perspective the situation becomes less clear. Recall that people need to do what they can to avoid being ignorant, powerless, or indifferent. Because of Leah's inexperience, she did not know how to evaluate problems with the instrument or the procedure for fixing it. Moreover, she did not look for spare filaments in case of breakage, so that she became unable to supply a replacement. In other words, Leah placed herself in a situation in which both her knowledge (of the instrument) and her freedom (to get a replacement filament) would be limited. If she did this with full knowledge, freedom, and approval, her responsibility becomes higher again.

The "Sainthood" and "Devil" Problems

So far we have spoken entirely about limiting moral responsibility. There is a flip side to this issue, which we might call the "sainthood problem." What happens when there are several morally good options? Do we need to pick the best one every time?

To answer this question, we will use the principle that people should try insofar as possible to continue to progress in the moral life (a major principle of Chapter 4). Such growth is difficult without continuing attempts to do better. Merely standing in place is difficult, and sooner or later usually results in backsliding. Hence, this principle suggests that we should try as much as we reasonably can to pursue the best option when several good ones present themselves. That is, if we choose a lesser option, there should be some good reason for doing so.

We must also confront the opposite of the sainthood problem, which we might call the "devil problem." What happens when we apparently have no morally good option? Fortunately, situations involving only "bad" options tend to be relatively rare, but when they do occur they produce some of the most wrenching decisions faced by humankind. Examples include the life-and-death stakes seen in hostage taking, on battlefields, and in lifeboats. To answer this question, we use the additional principle that the obligation to avoid harm outweighs the obligation to do good (a major principle of Chapter 6). Thus, we need to avoid harm as best we are able—meaning that our obligation to choose the least bad option is quite strong. In other words, there is less flexibility in choosing than there is when we have several good options.

A REAL-LIFE CASE: Responsibility in Software Engineering

Computers and their associated software affect a continually growing fraction of peoples' day-to-day lives. In applications ranging from video games to graphic design to safety assurance, software failures comprise an increasingly large fraction of system defects in comparison with hardware failures.

For example, during the period 1985–1987, six known incidents of massive radiation overdose occurred during use of Therac-25 medical accelerators for cancer treatment. The overdoses resulted in severe injury or death, and were caused by faulty control software. The software was designed to perform several functions, including keeping track of instrument status, accepting input for the desired treatment, operating key features of the instrument, and taking appropriate action in response to errors. The software was written over several years by one person (whose educational background surprisingly remains unknown to this day) who recycled a certain amount of code from earlier versions of the Therac. Documentation was poor. The manufacturer's fault tree analysis for the instrument as a whole did not include possible failures in the software.

Unlike more well-established disciplines like civil or mechanical engineering, software engineering lacks the professional organizational structures needed to set standards for definitions, recommended practices, or ethics. A lot of crucial software, even for life-critical systems, comes from small firms. While these firms provide fertile ground for the creative, freewheeling thought that has driven much of the industry's success, turnover is usually high and management oversight of engineering decisions is often low. Many observers agree that most accidents involving complex technology originate from a web of technical, organizational, managerial, and possibly sociological factors of which software may comprise only one part. Pressures to get a product to market quickly usually play an important role. Nevertheless, since software is often rewritten in many versions by many programmers whose contributions are poorly defined, assessing responsibility for system failure becomes very difficult.

◆ Have you personally ever encountered a bug in commercial software?

◆ What responsibility does a company have to fix bugs in its software? Does the application matter?

References

Boyer, Kevin W., *Ethics and Computing*. Los Alamitos, Calif.: IEEE Computer Society Press, 1996, 139 ff.

Levinson, Nancy G., and Clark S. Turner. "An Investigation of the Therac-25 Accidents." *Computer* 26 (1993):18–41.

" . . . to see what is right and not do it is cowardice."
CONFUCIUS (551–479 B.C.) *ANALECTS*, BOOK II, 24

Note

1. The authors are indebted to Edmund J. Seebauer, retired Director of Claims at State Farm Fire and Casualty Company, for providing the substance of this discussion.

Problems

1. Write a page or two describing an ethical dilemma involving moral responsibility you have encountered in a job you've had. (If you've been lucky enough never to have been confronted with a problem like this, describe one that a friend or relative of yours has had.)

   a. List the options/suboptions available to the main character who has to make a decision, together with the event tree flowing from each option. Indicate your estimate of the probability for each consequence on the tree (high, moderate, low).

b. Write down an expression for exterior goodness according to Eq. 6.1 as shown in the text example.

c. Identify a "responsibility issue" dealing with something bad the person did in the case or might have happened as a result of the persons's choice. Following the example in the text, list the factors that increase and/or decrease the persons's responsibility for the negative happening.

d. Recommend what you think the person should have done. Note: you don't have to say what was actually done in real life (unless you want to)!

2. Each case below has a question after it.

a. List the options/suboptions available to the main character who has to make a decision, together with the event tree flowing from each option. Indicate your estimate of the probability for each consequence on the tree (high, moderate, low).

b. Write down an expression for exterior goodness according to Eq. 6.1 as shown in the text example.

c. Each case has a "responsibility issue" listed after it dealing with something bad the character has done in the case or that may happen as a result of the character's choice. Following the example in the text, list the factors that increase and/or decrease the character's responsibility for the negative happening.

d. Recommend what you think the character should do.

CASE 8.1 Fear of Emotional Violence

Myra Weltschmerz picked up the telephone in her apartment nervously, and dialed the number of her boyfriend Martin Diesirae. She had promised him the day before to help him this afternoon with a term paper. He wanted her to put together some figures on the computer while he finished the text. Martin had been under a lot of stress lately with courses and with his job, so she had readily agreed. The trouble was that over lunch, Dolores Sola had called with an emergency babysitting request. Dolores was a hard-luck divorced mother of two for whom Myra baby-sat regularly. Dolores had accidentally driven through a stop sign and been hit by another car. Her car had been severely damaged, and she was trying to organize repairs. She had called and asked Myra to supervise the children when they returned from school. Myra had agreed, but she know Martin would not be happy.

"Hello?" came his voice.

"Hi, Martin. It's Myra. About this afternoon . . . something's come up." Then she related the story to him. He listened in silence. When she finished, she said, "Martin, Dolores has a real problem. I hope you understand. This won't cause you too much trouble, will it?"

There was yet more silence. This frightened Myra. Then Martin exploded. "It certainly will cause me trouble. Major trouble! My term paper is due at 5 p.m. today, and I'll never finish unless you help me with the figures! This is a huge part of my grade. You promised you'd be here!"

"But Martin, I thought your paper was due tomorrow," she ventured timidly. "I'll help you tonight, as long as it takes."

"No, Myra, I told you two days ago the due date was changed! Did you forget already?"

"I . . . I guess so. I'm sorry. . . ." she stammered.

"And I'm even sorrier," he ranted. "Tell Dolores to get someone else to baby-sit!"

"Martin, she was all upset! It's hard to organize baby-sitting from a repair shop."

"So let her take a cab home and finish what she has to do at the shop over the phone! I need you here, and you promised to help me."

"It's only for a couple of hours. . . ."

"I need those couple of hours," he burst in. "The one time I really need help, and you let me down. You whine and moan forever about how men break promises to you. But you don't keep your own! That's the last straw!"

Myra started to shake. She tried feebly to head off disaster. "Martin, I can't . . ."

"That's the story of your life!" Martin raged. "Can't, can't, can't! You can't feel happy, you can't handle being alone, you can't trust, and now you can't help me! And it's always someone else's fault. I am *so* tired of your wandering through life like this helpless little victim! All you worry about is your own pain, your own loss. You don't baby-sit for Dolores out of the goodness of your heart! You baby-sit because it makes you feel needed! I'm just tired of it!"

Martin's words ripped through Myra like bullets. She stood shocked and shaking, holding the phone limply. "Martin, please stop . . ." she begged weakly.

But Martin continued relentlessly. "No way! I don't have to take this any more—all the 'poor little me' stuff I endure from you!"

Myra had become numb, incapable of responding. Sensing this, Martin fired his final shot. "If you want to show me you really have a stake in this relationship, you'd better help me like you promised."

◆ What should Myra do? Responsibility issue: Myra's choice

CASE 8.2 Fear of Physical Danger

Martin Diesirae's eyes stung as he made his way through the smoke-filled hallway. Adrenaline surged through his body. A hot, hungry fire had struck the factory of Tripos Metal Polish, and Martin knew how dangerous that could be. With flammables and vats of fuming acid scattered throughout the factory, the fire could turn lethal at any moment. The fire alarm's deafening wail made his ears throb. He reached the factory offices and checked to make sure they were evacuated. To his surprise and dismay, he found his student coworker Emily Laborvincet in the accounting room. She was feverishly throwing computer disks and paper records into a box.

"Emily, what are you doing? There's a fire! Get out of here!" Martin shouted.

Emily glanced up briefly. "I'm going!" she responded, "but I've got to save these records."

Martin looked back up the hallway. The smoke was thickening rapidly. The fire alarm's insistent wail made thinking difficult. "Emily, you don't have time! Leave it! We've got to go!"

Emily remained determined. Not looking up from her work, she answered, "You go. There's a window in this office. I can get out that way!"

Martin couldn't believe his ears. "Emily, we're on the second floor. You can't just jump!"

"Sure I can. If I have to, I'll toss these out, then hang off the window sill and drop myself down."

Fear began to overtake Martin. "There's no telling what's down under the window! It could be burning already! Do you have any idea how bad this fire is?" he screamed. "We have to go now! The building could explode!"

"No way!" she responded. "There aren't enough solvents and flammables."

"But the fumes . . . acids, plastics, everything! They'll kill before you know it! And smoke! Look at it! I can hardly see down the hallway!"

Emily started to fill a second box. "Martin, these records are the lifeblood of the business. We can't replace them! They have to be saved. I just need one more minute. I can get out the window if things get too bad!"

Martin rushed over, grabbed Emily's arm, and started to pull her out of the room. "No, it's too dangerous! We have to go now, or we won't make it down the stairs!"

Emily wrenched her arm free in fury. "Don't touch me! You go if you want! No one is stopping you! Leave me alone!"

Martin rushed over to the window and tried to look out. But he couldn't see to the ground because smoke obscured the outside of the building. He glanced back down the hallway, and now saw flames. The stairway would be cut off in seconds. His heart beat wildly, and sweat poured down his forehead. He thought about overpowering Emily and carrying her down the stairs. It was now or never.

◆ What should Martin do? Responsibility issue: Martin's choice

CASE 8.3 Fear of Physical Violence

"I wish I hadn't forgotten my wallet," Emily Laborvincet muttered to herself as she picked her way around burned debris in the Tripos Metal Polish factory. "The one night I have to myself for shopping! Now I'll never make it to the store before it closes. I need my credit card to buy that new outfit. I must have left it in the top drawer of my desk." The darkness of the late evening cloaked the factory in weird, fantastic shadows. The recent fire had damaged the lighting system, and the regular lights were switched off after working hours. "I'll be out of here in a moment," she said to herself as she glanced about apprehensively. Emily got to her office door and pushed it open. She froze instantly upon seeing a large figure moving at the opposite end of the office.

Startled, the man whirled around, holding a small wad of cash in his hand. "Rolf!," Emily gasped. "What are you doing here? I thought everyone had gone home!" Her eyes settled on the money in his hand.

"I wasn't doin' nothin'. I don't like it when people surprise me," he growled.

He took a step toward her, and Emily caught sight of the petty cash box the company kept for small expenses sitting open on the table behind him. "Rolf! What are you doing with the petty cash? You're in production! You're not supposed to be in there!"

"What's it to you?" he growled again. "I was just makin' a little change for myself."

As the company's accountant, Emily suddenly remembered that small amounts of cash had been disappearing from the box for several weeks. It was never more than thirty dollars at a time. Since she'd been so busy, especially since the fire, Emily had chalked the shortage up to bad records. Now she suspected differently. "You haven't been stealing out of there, have you?" she asked guardedly.

Rolf moved yet a step closer. "I told you, I was makin' some change for myself. There's no harm in that, is there?" He gave her a long look up and down.

Emily unconsciously backed away a step and glanced behind her.

Her heart beat faster. "No, Rolf, I was just asking. There's been some cash missing from there. I just wanted to know where it went."

Rolf drew himself up to his full 6-foot 4-inch, 240-pound size. He had played linebacker fifteen years ago in high school. He licked his lips and scanned Emily again. "I don't like it when people call me a thief," he snarled. "It's not polite." He drew yet a step closer, only 6 feet away now.

Sweat broke out on Emily's forehead, and her voice grew breathy. "I didn't mean to insult you, Rolf!" She glanced behind herself again. "I'm just trying to do my job." She looked down at her watch. "Oh, look at the time," she chuckled weakly. "My boyfriend's expecting me. We're having a late supper. He'll be mad if I'm late." She turned quickly to go.

"Wait!" commanded Rolf. She froze. He walked up, grabbed her arm, and whirled her around. "You're not going to say nothin' about this, are you?" he demanded threateningly.

She shook her head stiffly. "What's there to say? You were making change."

He glared at her for a moment, and leaned over so she could feel his hot breath on her face. "Good. I'd better not catch you sayin' nothin' to nobody." He looked her over yet again with a malicious grin. "I'm sure your boyfriend likes that pretty face of yours. It sure would be a shame if anything happened to it, wouldn't it?" Emily nodded slightly. "Now get outta here," he ordered. She turned to go.

◆ Should Emily say anything about the incident to anyone? Responsibility issue: Emily's lie to Rolf

CASE 8.4 Distributing Profits among Partners

"Thanks for helping me with all this programming," said Leah Nonlibet to her boyfriend Terence Nonliquet. "Professor Clark said he was really thankful. He said now we should be able to do the data analysis he needs for the meeting he's going to in a couple of weeks."

Terence looked up from the computer screen in front of him and frowned. "Well, I'm glad I could help. But I still say it was bad planning on Clark's part. He should have had one of his graduate students working on this long ago. Instead, he asks you, an undergraduate hourly worker, to do it over your Christmas break."

"He was desperate," Leah replied. "I told you how disorganized he gets sometimes. He forgot that most of his graduate students were gone over the holidays. I was staying in town for most of the break, and he didn't have anyone else to turn to."

"I can't wait till that meeting is over," Terence huffed. "First Bernard guilts you into drawing the graphs for his presentation there, just because you accidentally broke his mass spectrometer. And now Clark begs you to do the data analysis for *his* presentation."

"Like I said," Leah insisted, "Professor Clark didn't have anyone else to turn to."

"But you couldn't do the job either," Terence exclaimed with irritation. "The code you're supposed to modify is in C. You know only Fortran. Didn't you tell him that?"

"Yeah, but he said the two weren't that different. He said he thought I could just pick up what I needed to know as I went along. Who was I to question him?"

"So I had to bail you out when you ran into trouble," Terence grumbled. "You know, Leah, I was hoping to get some rest over the break. Last semester was just *so* busy, and things won't be any better in the spring."

"I know, Terence. And I'm grateful—really." She sashayed over and kissed him on the forehead. Terence smiled weakly.

"So what kind of commission am I going to get?" he inquired vaguely, returning to his computer.

"Commission? Oh, come on Terence!" Leah laughed. "You're so funny sometimes."

Terence stopped working and eyeballed her. "Actually, I wasn't totally joking," he retorted gravely. "I mean, your gratitude is nice, but shouldn't there be something more? After all, Clark is paying you by the hour, but I'm doing a big part of the work."

Leah tensed. "Well, I don't know. I never thought about it. I just thought it was something a boyfriend would do for someone he cares about," she offered.

Terence's voice grew sharper. "Who told you that? You're right, I'm happy to help you. But fairness operates even between boyfriends and girlfriends. When I do work you're getting paid for, I think you owe me part of the profits."

"What part?" Leah demanded incredulously. "I put in a lot more hours than you—you have to admit that! And I told you I appreciated what you did."

"Yeah, but sometimes it would be nice to get appreciation that I can put in my wallet. Sheeze, Leah. I don't know why you're so shocked. I mean, when we go out to movies and plays, I pay. When we go out to eat, I pay. It's not like I don't take care of you. I hold doors open for you, get you flowers, all that kind of stuff. Now I ask for fair payment for work I do for you, and you look at me like I'm crazy!"

"But I pay for popcorn at movies, and tips at restaurants," Leah rejoined defensively. "So you don't pay everything. Anyway, in my family it was always considered gentlemanly to do what you do. My brothers do it for their girlfriends. And since you just started doing it when we started dating, I thought you felt the same way." Leah paused to reflect, then continued. "I know some couples split everything, fifty-fifty. Every movie, every meal—everything. I guess that's OK, but it seems so businesslike. Sort of like a contract. But I like to think of dating as more than a business contract. A lot of guys feel that way too. The guy I dated before you—he spent every last dime he had on me!" She sighed. Then she grew stern. "But I can't be bought. He found that out!"

"Leah, it's easy to think that way when you're not footing the bills," Terence remarked impatiently. "I don't have the kind of money to blow that your last boyfriend did. In fact, I don't make much more money than you. You obviously like it when guys lavish you with stuff, but then you say you can't be bought. That's contradictory—you can't have it both ways." Terence's voice turned bitter. "I've met other women like you. When you set some kind of bar for how much a guy has to spend on you, you set a price on yourself. You may want trust, commitment, and all those things, but you set a price all the same. If he doesn't pay it, he's history."

"Terence, I'm not like that!" Leah gasped.

"No woman I've ever met thinks she is," he shot back. "Look, let's get back to the main point. You couldn't have done this work for Clark without me. I put in about a third as much time as you did. So out of the total hours, I did a quarter of them. I'm not obligated to work for free, even if I'm subcontracting for my girlfriend. So I want a quarter of what he pays you. That seems fair to me."

◆ What, if anything, should Leah give Terence? Responsibility issue: Leah's choice

SUMMARY

The main principles and methods for approaching a complex ethical case may be summarized as follows:

1. List who is involved and what interest each has in the case.
2. List important circumstances regarding person, time, and place.
3. Devise options for action (with suboptions where appropriate).
4. Create an event tree of consequences for each option.
5. Show on the tree your qualitative estimate for the likelihood of each consequence.
6. Rate the goodness of each consequence by deciding how the consequence accords with the virtues. Rate consequences that reflect justice, prudence, temperance, and fortitude as "good," and those that do not as "bad."
7. Balance the exterior goodness of the consequences by summing all the consequences using:

$$\text{Net goodness} = \Sigma \text{ (goodness of each consequence)} \times \text{(importance)} \times \text{(likelihood)}$$

8. Make educated guesses about the interior attitudes toward exterior consequences and suggest how such attitudes will imprint the tendency to follow the virtues.
9. Balance exterior and interior goodness, attending to how many people the effects of interior imprinting will touch.
10. Pick the option with the best mix of exterior and interior goodness. When two roughly equal "best" options appear, choose the one whose individual consequences seem least bad.

Some Words of Caution

The method we have laid out in this unit may seem relatively rational and tidy. However, great dangers await those who apply it blindly. Failures in prudence can cause us to unconsciously ignore unpleasant consequences, leading us to leave important terms out of the equations. Failures in temperance or fortitude can cause us to exaggerate the moral importance of some consequences. Failures in justice are most poisonous of all, and cause us to pick options that are simply bad.

Few practices can better preserve us from failures in virtue more reliably than regular, honest assessments of our intentions. Besides helping us to pursue virtue, looking at our intentions regularly can help remind us that many human actions come from mixed motivations. Keeping this fact in mind helps to reduce the human tendency to demonize those we don't like by assuming they act with only evil intentions.

UNIT THREE

JUSTICE: APPLICATIONS

"By a lie a man throws away and . . .
annihilates his dignity as a man."
IMMANUEL KANT (1724–1804), *DOCTRINE OF VIRTUE*

"Let justice be done, though the sky falls."
WILLIAM MURRAY, EARL OF MANSFIELD AND LORD CHIEF JUSTICE
OF ENGLAND (1704–1793), *REX V. WILKES*, 1768

9

TRUTH: PERSON-TO-PERSON

"Whoever is careless with the truth in small matters
cannot be trusted with important matters."

ALBERT EINSTEIN (1879–1955), *IDEAS AND OPINIONS OF ALBERT EINSTEIN*

Many ethical concerns focus on the virtue of justice. Justice has two aspects: truth and fairness. The present chapter focuses on truth as it operates on a person-to-person basis. By "person-to-person" we mean situations in which relatively few people are involved and who know each other.

Truth in Actions

Consider this example. Suppose it's the December holiday season, and you attend a family gathering. Before going, your mother reminds you to spend some time with your weird Uncle Harold. "Be sure to tell Uncle Harold about your semester at college. He likes you so much, you know. And he's so lonely—never having married, and all. It's important to be nice to an old man like that, and a relative on top!" Right away you imagine what will happen. Uncle Harold will light up one of those horrible cheap cigars he always smokes and listen to your summary, sometimes laughing at things you never intended to be funny. Then he will tell some story from his youth for the hundredth time, and will offer all sorts of advice on how to choose a marriage partner. By this point, you will be ready to bolt from the room screaming. However, weird Uncle Harold is rich. Before you excuse yourself, he will slip you a crisp one-hundred-dollar bill, winking and whispering that you should keep this "between the two of us." So, while you realize that Mom is right in what she says about Uncle Harold, you also remember that you really want the money to help buy a new bicycle. So you respond, "OK, Mom," and she pats your shoulder for valuing the family's peace so highly.

Have you acted in truth? A devil's advocate might claim you're being a hypocrite—paying lip service to one ideal but pursuing another. In fact, your motivations are *not* as single-minded as either your mother or Uncle Harold might think. It's not that family togetherness and kindness to old men aren't important, but that you can really use the money. Clearly, the truth of your action depends heavily upon your intention. Since human intentions can be mixed, the question of truth becomes tricky to resolve and may not submit to a black-and-white answer. Nevertheless, the issue remains very important. For some people the practice of hypocrisy becomes so typical that we say they're "living a lie."

To project the true and right impression, we need to communicate our intentions verbally where necessary to explain what we do. Unfortunately, few concrete rules exist for when or how to do this. Supplementing actions with words is an important part of the craft of ethics. As with any craft, skill comes mainly with experience.

Truth in Words

The last example dealt with the truth of an action: being nice to weird Uncle Harold. The question of truth revolved around whether your intention actually matched what it appeared to be from the outside. Now consider a simple case that focuses more on the meaning of words. Suppose you maintain your home address with your parents in California but attend college in Wisconsin. Suppose further that you want to buy a high-powered hand calculator through a discount mail-order supply firm. The order form says, "Residents of CA and WV must add state sales tax." On the order form, you list your college address in Wisconsin and add no tax. Have you told the truth? Since your legal residence remains in California, a devil's advocate could argue that using your temporary Wisconsin address amounts to lying. On the other hand, you might respond that you spend most of your time in Wisconsin, so that you could truthfully claim to live there.

Here, truth revolves around how we choose to define the word "resident." The state of California might prefer a definition based on legal residence, whereas you might prefer one based on where you spend most of your time. Put differently, truth can become very muddy where words having multiple meanings become involved. Once again, to project the true and right impression, we sometimes need to indicate clearly exactly what we mean by the words we use.

Harm from Deception

So far we have purposely discussed cases with shades of gray to point out where trickier issues of truth crop up. However, certain words and actions clearly and willfully project something false, either directly or through their context. We refer to such behavior as *lying,* or if intended to ruin another person's reputation, as *slander.*

What happens when we lie? Sometimes lying can help bring about good things or prevent bad ones, at least in the short term. However, in the longer term bad consequences often appear that are very severe, including:

1. other people will no longer trust you if the lie is discovered
2. you may need to tell more lies to cover up the first one
3. you may unexpectedly mislead others into trouble

Sometimes these things seem unlikely to happen, making the temptation to lie very strong. Nevertheless, the idea of character virtue points to another important problem: lying once makes it easier to lie again.

In the practice of science or engineering, lying can cause especially nasty problems. Misrepresenting data or procedures to your coworkers strikes at the core of an effort intended to discover truth about nature. Inflating your own contribution to the effort at the expense of others corrodes the team effort needed for long-term success.

Are there any circumstances under which lying can be accepted? Most moral traditions look upon lying with *extreme* suspicion, and several important ones forbid the practice altogether under any circumstances.[1] The question is a complicated and difficult one, and in almost all practical cases is beside the point. In nearly all professional practice, lying cannot be justified, and honesty brings about the greatest long-term benefit. Only in situations involving extremely serious consequences (like threats to life) and minor falsehoods might lying be allowed. However, we emphasize that such cases are very rare.

Harm from Withholding Truth

Of course, outright lying is only one way to deceive. *Mental reservation* involves withholding important information without actually lying. People commonly use mental reservation to answer a question they would prefer not to touch. They tell a partial story instead of the complete one. Since no outright lying is involved, moralists consider mental reservation to be acceptable under some circumstances.

In many cases, however, withholding what others need to know can cause many of the same problems as lying. While many people rank improperly withholding truth as a lesser offense than lying, we still need to view the practice with suspicion. Note that withholding truth does not necessarily require silence. In fact, in science and engineering where a lot of information and data exchange takes place, truth can be withheld by burying it under mountains of information—as a small phrase tucked into a long memo or as a single entry hidden in a group of bulky tables. Inappropriate withholding of truth becomes especially dangerous in situations involving safety or quality control. When a production run is pressed to meet a deadline, the one who calls for a halt because of bad safety practices or corner-cutting on quality usually makes few friends by doing so. Short-term worries about meeting deadlines can easily obscure longer-term threats to worker health or customer satisfaction. Big sums of money can be at stake. It can become very difficult to decide when silence is prudent—specific cases usually require balancing the short-term gains against the longer-term harms.

Problems with withholding truth sometimes arise in disputes over contracts. This issue is assuming increasing importance in the United States, where in some markets the costs of directly owning equipment and expertise are increasing rapidly. These costs are fueling a trend for companies to contract work out to specialized firms rather than do it in-house. Sometimes problems come up that are not specifically addressed in the written contract. If no one expected the problem to occur and everyone negotiates in good faith, there is no ethical breach. The situation changes, however, if someone expected the problem without saying anything, or tries to negotiate maliciously.

Whistleblowing

Occasionally you may see a coworker doing something wrong in a situation where the moral stakes are very high. For example, you may observe (or be asked to participate in) flagrant offenses against safety, environmental standards, work-place equality, and the like. Sometimes these offenses take place with the unstated or even explicit approval of management. To stop them would require you to appeal to very senior levels of management or to oversight agencies outside your organization. Such an appeal is commonly termed "whistleblowing."

Whistleblowing presents a painfully difficult moral choice. Commonly the options boil down to "should I say something to stop these abuses and risk severe retaliation, or should I remain silent and stay out of trouble?" Several studies of whistleblowing confirm the common sense conclusion that whistleblowers generally face hostility within their organization and commonly leave their jobs, either voluntarily or by firing.[2,3] Extended litigation is often involved. On the other hand, remaining silent while severe abuses continue does not stop the injustice and eats continually at the conscience.

It is true that a whistleblower who is fired can often find another satisfactory job. In fact, some employers greatly prize workers who demonstrate commitment to high ethical standards even at great personal cost.

Nevertheless, no such job is guaranteed, especially in a tight market. A newly unemployed whistleblower may need to relocate, and may lose seniority and retirement benefits. Moreover, whistleblowers sometimes show little of the team spirit that many employers value, instead coming across as complainers or fanatics.

Unfortunately, few general rules exist to promote or discourage whistleblowing, and most situation must be considered on a case-by-case basis. Most moralists agree that because of the potential severity of the retaliation, whistleblowing is praiseworthy but is not required under all circumstances. However, one standard does apply: the duty to blow the whistle increases as the gravity of the wrongdoing increases.

Harm from Spreading Truth

While it may seem obvious that the pursuit of truth discourages hiding many matters, it may seem less obvious that this pursuit sometimes also discourages spreading those matters unnecessarily.

Gossip and *scandal* involve needlessly spreading damaging truths about others. "Gossip" suggests idle chitchat about minor things, while "scandal" implies more gravity and malice.[4] Both gossip and scandal deal with true facts (as opposed to slander, which contains falsehood), but involve either indifference or malice toward another's reputation. Hence, gossip and slander are unethical because of the poor intentions. Notice that if the same damaging facts are shared with good intentions with appropriate people, the situation changes completely. In fact, the action could change from very bad to very good!

In science and engineering, one other practice connected with truthtelling deserves special mention: *disclosure of trade secrets*. Inventors, by securing a patent, can enjoy the exclusive right to use or sell what they have created. However, patents have little practical value if infringement cannot be proven easily or cheaply. Hence, many companies protect their inventions simply by keeping them secret. This strategy works well as long as employees remain loyal and as long as those who leave the company do not tell what they know. Inappropriately disclosing a trade secret is plainly unethical. Companies often try to ensure silence by asking departing employees to sign nondisclosure agreements or agreements not to work for competitors for some length of time.

Privacy

"Privacy" is a word that carries many meanings. Privacy can refer to a person's freedom to be alone, freedom to act without intrusion of others, and freedom to control what others know about him or her. In the context of truth, we will concern ourselves mainly with the last meaning. Before dis-

cussing ethical issues surrounding privacy, however, we need to ac-
knowledge that people's conception of privacy depends heavily on culture and upbringing. For example, citizens of the United States often tend
to value privacy more than citizens of Far Eastern countries. We will not
explore the justification for privacy here, although such justifications play
key roles in determining how far privacy rights should extend. Instead, we
will simply point out where ethical issues concerning privacy tend to come
up in science and engineering.

One such arena concerns information technology for personal data.[5]
Such data include medical histories, credit card usage, financial records,
education records, and the like. Computer scientists and engineers in particular can find themselves on either side of this problem. They may become involved with compiling and protecting such databases from
unauthorized use. Since increasing security usually decreases ease of access, the ethical question generally involves balancing the risk of security
breach against the difficulty of access. (On a more social level, there may
also be questions about legal but questionable uses, such as selling personal data merely for marketing purposes.) From the reverse perspective,
employees may find themselves in a position of easy access to personal
data. Or more aggressively, "hackers" may attempt to break into secure
systems. Ethical questions come up even if all the intruder does is look,
rather than change data or deposit a virus.

Another arena for privacy issues that affects a larger cross section of
scientists and engineers is surveillance in the work place by employers. In
the United States, employers are watching their employees ever more
closely by monitoring telephone conversations, computer keystrokes, Internet site visits, and e-mail correspondence. Furthermore, unannounced
tests for illegal drugs have become common. Here, the main ethical issue
involves balancing the need for an employer to ensure productive activity
against the right of an employee to protect certain personal activities from
observation. An important standard for striking this balance is that the right
to invade privacy (for the good of the innocent) increases with the gravity of the suspected wrongdoing.

A REAL-LIFE CASE: Censorship of the Internet

Free communication protects many ideas that the virtue of truth em-
bodies. However, as we have pointed out, sometimes indiscriminate
broadcasting of information does more harm than good. This idea
lies at the heart of the ongoing debate about censoring content on
the Internet that is obscene or pornographic. Before the Internet age,
when books, newspapers, radio, and television were the primary

means of public communication, U.S. law had found a workable compromise between the need to preserve easy communication among adults and the need to protect children from seeing or hearing things they were not ready to digest. Hence, there were limitations on what could be heard or seen on radio and television (at least on certain channels during certain hours), and bookstores needed to ensure that their merchandise was displayed so that children could not access certain material.

Such compromises have proven much more difficult to arrange for the Internet. Several factors contribute to the problem. First, the Internet is so large and diffuse that monitoring and control are nearly impossible. International cooperation would be required because of foreign Web sites. Second, difficult technical challenges accompany attempting to restrict access to certain sites by certain people. While so-called blocking software can be installed on a computer to stop access to certain sites, the effectiveness of such software is questionable. Third, parental control over what their children see can be difficult to enforce because the children often know more about the Internet than their parents, and because parents can control computers only within their own homes.

In an attempt to solve the problem, the U.S. Congress enacted and the President signed the Communications Decency Act in 1996. The law banned Web sites containing certain "indecent" words and imposed prison terms and stiff fines as penalties. However, the crudeness of the criteria for "indecent" material in effect unintentionally outlawed sites offering classic literature, education regarding sexual practices, and bulletin boards regarding breast cancer. The U.S. Supreme Court struck down the law the following year as an unconstitutional infringement of free speech, but the issue remains alive in Congress.

◆ Do you agree with the Supreme Court's decision? Why?

◆ To what extent should the burden of protecting children from obscene material fall on parents as opposed to the society at large?

◆ What, if anything, should the government be doing to solve the problem?

References

Beannefield, Robin M. "When Kids Prowl the Net, Parents Need to Be on Guard." *U.S. News and World Report*, 29 April 1996, 75.

Novak, V., and C. Stamper. "Free Speech for the Net," *Time*, 24 June 1996, 56.

"The liar's punishment is, not in the least that he is not believed,
but that he cannot believe anyone else."

GEORGE BERNARD SHAW (1856–1950), *THE QUINTESSENCE OF IBSENISM*, CH. 4

Notes

1. One philosophical tradition (deontology), represented by the writings of Immanuel Kant (1700s), forbids lying altogether. However, another tradition (natural law), represented by the writings of Thomas Aquinas (1200s), permits minor lying in very rare and exceptional circumstances that bring about extraordinary good or prevent grave evils and that are required by prudence.

2. In the academic research establishment, about 12 percent of whistleblowers who face serious negative consequences ultimately lose their jobs—according to a 1995 survey conducted by the U.S. Department of Health and Human Services. However, the most common response to whistleblowing is pressure to drop the charges, followed by countercharges. See Sarah Glazer, "Dealing with Threats to Whistleblowers," *CQ Researcher* 7 (1997):6.

3. See Mark Lasswell and Joseph Harmes, "Justice Deferred," *People*, 12 June 1995, 101–102, for a compelling example case in which a whistleblower was harassed and fired by his employer. Eventually the whistleblower sued and was awarded $13.7 million.

4. "Scandal" has another meaning that may be more familiar: any disgraceful action or circumstance, usually having a public dimension.

5. For a more complete discussion, see John Weckert and Douglas Adeney, *Computer and Information Ethics* (Westport, Conn.: Greenwood Press, 1997), 75 ff.

Problems

1. Write a page or two describing an ethical dilemma involving truth you have encountered in a job you've had. (If you've been lucky enough never to have been confronted with a problem like this, describe one that a friend or relative of yours has had.) Recommend what action you think you (or your friend/relative) should have taken, and give reasons for and against that recommendation. Note: you don't have to say what was actually done in real life (unless you want to)!

2. Each case below has a question after it.

   a. List the options available to the main character who has to make a decision, together with the event tree flowing from each option.

   b. Recommend what you think the character should do.

CASE 9.1 **Full Disclosure of Contract Terms**

"So that's the story," Dolores Sola sighed to Myra Weltschmerz. Dolores glanced at the clock on the kitchen wall in her apartment. "Eleven thirty," she moaned. "Even if I go to bed now, it'll be two hours before I can sleep. Then I have to get up tomorrow to get the kids off to school. I'll be a zombie."

"The important thing is that you didn't get hurt too badly in the car accident. Just a few sprains," Myra offered sympathetically.

"I'm sorry to keep you here so late!" Dolores exclaimed. "I had no one else to turn to for looking after my kids. I never dreamed the doctors would take so long to fix me up. And what a day you had . . . with your boyfriend blowing up on you like that! I'm so sorry it was on my account—it makes me feel even more miserable." The two sat in silence for a moment. "So do you really think you and Martin are through?"

Myra shrugged. "I don't know," she replied forlornly. "I'm too numb to think about it. It's such a shock. Let's not talk about it tonight."

Dolores straightened up in her chair. "Well, we can talk about it tomorrow if you like. I'll see you here when I get home from work."

Myra jolted out of her sad musing. "Oh, Dolores, I can't baby-sit tomorrow. I'm too much of a mess. And I have to study for a test on Friday. I was going to start tonight, but I didn't expect to be here so long. Now I have to do everything tomorrow evening."

Dolores's voice quickly turned plaintive. "But Myra, you always watch my kids on Tuesdays and Thursdays. Tomorrow's Thursday. I can't get home from work until 7—probably later, since my car's in the shop, and I have to take the bus. And I've already missed a lot of days at work. If I miss any more, I think I'll get fired."

"I'm sorry, Dolores, you'll have to get someone else. I just can't."

Dolores grew more insistent. "I can't get someone else! I have other sitters, but they've all told me Thursday nights are out. You're the only one I've got. And you know I can't leave Garrett and Lorelei alone for long. They're only 10 and 8, and always get into trouble. Anyway, we had an agreement. You can't back out now!"

"What agreement?" Myra demanded. "It's not like we ever signed anything. I sit for you on Tuesdays and Thursdays because you ask me to every week. You could stop using me at any time. I always assumed I could get out of it if I needed to. You never told me I was your only sitter on Thursdays!"

"You must have forgotten," Dolores remarked. "I'm sure I mentioned it sometime in September." Her voice turned angry. "Anyway, you've always done it—never missed. Since last August! That's at least an implied contract. And I pay you more than my other sitters because you're so reliable. I know I told you that too!"

Myra wracked her memory, trying to remember these conversations. "I don't remember all the things you've said," Myra retorted, voice quavering. "But using me as your only sitter wasn't smart. What would you do if I got sick? Now I'm an emotional wreck and I've got an exam coming. That's a good reason to miss. It's your fault if you didn't round up a backup for me."

Dolores burst into tears. "Oh, fine. Run out on me!," she cried. "Everyone else always does! Countless men . . . employers . . . everyone! I thought you were different! I thought you understood! If my kids kill themselves, or if I lose my job, it's your fault!"

◆ What should Myra do?

Case 9.2 Overhearing Telephone Conversations

Leah Nonlibet shuffled up slowly to the closed door of her boyfriend Terence Nonliquet's office. The spring semester had started, and he had resumed his duties as an undergraduate teaching assistant. Because of their busy schedules, Leah and Terence had agreed the previous semester to study together regularly, so they could see more of each other.

Still, a lot of tension persisted between them. Still unresolved was his complaint about her neediness. And she wanted to know what was going on between Terence and Celia Peccavi, a student in his class last semester. She had intercepted a letter Celia wrote him inviting him to a party. She had hidden the letter and dragged him off to a movie the night of the party, but apprehension burned hot inside her.

As she drew closer to the door, she could hear Terence talking. He was evidently on the phone. She paused, then put her ear up to the door. She could just barely make out his words. "So do you understand now Celia? Yes, that's it . . . you've got it. OK? Anything else? Call again if anything else comes up. Bye." Suddenly Leah could feel rage boil up inside her. She fought it back, and rapped sharply on the door.

"Come in . . . "

She yanked the door open. "Hello, Terence," she said stiffly.

"Hi, Leah. Cold out, huh?"

"Yes," she replied icily, removing her coat.

"Lots of work to do tonight."

"Yes. Who was that on the phone?"

"Just now? A student in my class. She needed some help on the homework."

Leah sat down on a chair at a distance from Terence, and crossed her arms. "Does this student call often?" she inquired in a measured tone.

Terence, oblivious to Leah's state of mind up to now, suddenly sensed trouble. "I don't know," he answered guardedly. "It's only the beginning of the semester. There hasn't been much time to talk to people."

Leah continued her thinly veiled interrogation. "So how well do you get to know your students?" she asked.

"Oh, I don't know. Some more, some less. It depends."

"Do you have any from last semester in your class now?"

"A few. Three, I think."

"So they must know you pretty well," Leah observed. "You're only a junior, so you're not too far from them in age. Do they invite you to things? You know, parties and stuff?"

Terence tensed visibly and glanced at the wall. "I don't know. I guess it probably happened once or twice. I don't remember for sure. No big deal."

Leah leaned forward in her chair, stonefaced. "Did you go?"

Terence's mind raced as he searched for a way out of the tightening noose. "Leah, I don't know . . . what difference does it make? Why are you asking me all this?"

Leah's eyebrows raised. "You don't know if you went?" She sat back in her chair and recrossed her arms. "I'm just wondering because of what we talked about. How we're so pressed for time we can't always see each other as much as we want. I thought that if you were going out to these social things, maybe I could go too."

"I doubt that," he muttered reflexively. He thought of the volleyball game he had attended at Celia's request, and the party that she invited him to several weeks later by letter. The letter had gotten misplaced (he didn't know how) until after the party, so he didn't go. But she had followed up by asking him for a date with pressure verging on extortion. He had resisted by not responding. Still, Terence could see little benefit to sharing all this with Leah.

"You doubt that!" Leah exclaimed. "Well, this is news! You're usually a guy who's pretty above-board . . . like when you recommended that software package to your class when the course instructor ordered you to recommend another one. Maybe there's some girl hitting on you?"

Terence squirmed in his seat. "W-why do you think that? I'm loyal to you!" he stammered.

Leah continued relentlessly. "So you're loyal. Good. You can't be too careful these days, you know. Somebody's always got an angle. Take that girl you said just called, for example. Maybe she's got some ideas."

Terence's head felt light, and butterflies beat madly in his stomach. "What ideas?" he said blankly.

"Maybe she's trying to make a move on you. Would you be interested if she did?"

"Hang on, Leah," offered Terence as he stood up. "I have to make a run to the bathroom." On his way down the hallway, he pondered his answer.

◆ What should Terence say?

CASE 9.3 Group Efforts on Homework

"You wanted to see me?" asked Celia Peccavi as she sashayed up after class to Terence Nonliquet, the teaching assistant for Comp Sci 110. "The homework set you handed back had a note on it."

"Uh-huh." responded Terence, shuffling some notes. "You and Jacklyn."

"Jacklyn's not here today," she responded mischievously. "So you and I can be alone."

"Celia, stop it!" Terence shot back impatiently. "I'm serious. I think the two of you copied from each other on the last homework assignment. If you did, I'm going to dock you points."

"No need to get upset," Celia purred. "I didn't copy. But I told you that if you didn't go out with me just once over last Christmas break, I would enroll in your class. You didn't, so here I am! True to my word." Her voice turned artificially wounded. "You could have at least called, you know. I just wanted you to give me a chance, but you ignored me."

"Celia, that's got nothing to do with what I'm talking about," Terence replied with exasperation. He took the paper out of her hand and pointed to it. "Look at this programming problem. Everyone was supposed to do their own work. But in grading, it was pretty clear that people were working in groups. I could tell from the approach and style people took in their solution. That's OK, as long as people showed some evidence of doing their own work." Terence's voice started to rise. "But it was different with you and Jacklyn. Your code has big sections that match each other verbatim. I mean, comma for comma, period for period! You use identical variable names. And your output graphs have exactly the same format—border widths, legend placement, almost everything! That couldn't happen by chance."

"Jacklyn and I did work together some," Celia acknowledged, "but we didn't copy. We did our own graphs and programming. See? We put different titles on the graphs." She pointed to the paper. "And check out this section of the code. And this one. They don't match exactly. And the comment statements are totally different everywhere."

"Well, I think what happened is that one of you wrote the main body of code, and the other copied the file. Then each of you played

around with a few details, and overlaid your own comment statements," Terence persisted. "In my book, that's cheating."

"No it's not—Jacklyn and I just think alike," Celia responded flip-
pantly. "You yourself said people could work in groups if we wanted.
What do you expect? That everyone will take a completely different
approach?"

Terence grew agitated. "No. But I expect to see some evidence
that the work came out of your own head!"

Celia smiled impishly. "But look at the comment statements," she
insisted. "They're obviously out of my head. I don't think your evidence that we cheated is very strong. There are only so many ways
to program this problem, you know."

Terence's face reddened. "Celia, no! This couldn't happen by
chance! You're not taking this seriously. If you can't convince me this
was your work, I'm going to dock a lot of points!"

"You're cute when you're mad!" she giggled. Then she put on a
pretended air of seriousness. "Terence, I didn't copy. And I don't
think you can judge whether I did without talking to Jacklyn. It's not
fair. You have to give us both the chance to defend ourselves." Then
Celia drew a step closer and eyeballed Terence. Her voice sharpened. "And if you take off points, I'll complain to my uncle. He's the
department head, you remember. Your boss's boss. And the one who
thinks the sun rises and sets on his sweet little niece. I don't think
he'll be happy that you're harassing her."

Terence threw up his hands in a huff, and tossed the paper on
the desk in front of him. "I don't know what to do with you!" he exclaimed.

◆ What should Terence do?

CASE 9.4 Disclosing Trade Secrets after Jumping Jobs

"So that's the story, Monica," said Emily Laborvincet. "A lot has happened since you and Chase got in that car wreck. It's enough that
Martin and I have been trying to fill in for you as part-time workers,
and undergraduates at that! But then half the factory gets wrecked in
a fire. You must be crushed! You worked so hard to build that business."

Monica Ichdien shifted herself in her hospital bed a little—as much
as the traction harnesses would permit. She sighed. "Yes, Emily, it's
hard. But don't worry. I've endured tough times before. My husband
died twenty years ago, after we were married for five. Bone cancer.
I suppose it was lucky we had no children yet. I learned a lot while
he lay dying—about what is important in life, and about how to live."
She smiled at Emily. "The business will come back. It's just a tem-

porary setback. And I have a lot of faith in you—and Martin and the rest—to carry on while I'm gone."

"But Monica, I'm only a sophomore. I'm still not sure even who I am, or what I want to do with my life."

"Look at it like this," Monica began. "If you take the situation at Tripos and make it work, there's no limit to what you can do. It will be a great learning experience and confidence builder. And such a unique experience will give you insight that most people work years to achieve. In the long run, a job like this can help you make better choices about your future."

"I never thought of it that way," replied Emily. She stopped to ponder. "Although it is coming at a big cost. I've had to pretty much give up my violin playing now . . . just when I was really starting to improve." Then, noticing Monica's strength beginning to slip away, Emily added, "Monica, there's one other thing. You remember that I worked last summer for Beta Chemical? They make a couple of etchants related to ones we make. Well, yesterday I was talking with Conrad, the lab chemist, about our process for making one of them. As he described it to me, I recalled that Beta used a special procedure to cut out the boiling step, saving a lot of money in heating costs. The trouble is, while their procedure isn't patented, it's a trade secret. They try very hard to keep it quiet."

"But it's for a different product?" asked Monica.

"Yeah, so we're not direct competitors."

"Didn't you sign a nondisclosure agreement when you left Beta?"

"I was supposed to, but things got confused at the end by accident, and it turns out I never did. You know Monica, with Tripos in such bad shape now, we could use anything that would make some extra profit," Emily pointed out earnestly. "I could tell Conrad how to do it."

"I'm not sure, Emily. I have to think about it when I'm not so tired."

◆ Should Emily disclose the secret?

10

TRUTH: SOCIAL

"Seek the truth
Listen to the truth
Teach the truth
Love the truth
Abide by the truth
And defend the truth
Unto death."

JOHN HUS (C. 1373–BURNED AT THE STAKE 1415)

In this chapter, we consider truth from a broad social perspective. By "social" we mean situations involving large numbers of people who know very little about each other.

Distinctions between Science and Engineering

Scientists pursue the truth of general physical law, attempting to uncover underlying principles that govern the natural world. While practical applications may help motivate the work, gaining knowledge for its own sake forms an important goal. Science is a social endeavor, with humanity as a whole sharing in its fruits.[1] Hence, the free interchange of data and ideas forms a core value of science, and the scientist must attempt to share methods and data as completely and accurately as possible.

Engineers, however, pursue the truth of practical applications, applying known principles to invent better substances, devices, or processes. Engineers concern themselves more with using knowledge than with gaining knowledge for its own sake. While engineering is also a social endeavor, the fruits need not be held in common by humanity as a whole. The engineer (or employer) often wants to claim some direct benefit from what was discovered, and may need to keep methods and data within the organization.

157

Of course, these distinctions are extremely oversimplified. No clear boundary separates science and engineering. Indeed, words like "applied science" and "engineering science" find widespread use. Furthermore, many scientists in the corporate world submit routinely to restrictions on publication, while many engineers in the university environment publish as freely as they like. Despite these inadequacies in our description, however, the fact remains that we can take two distinct approaches to knowledge about the physical world. One perspective seeks knowledge mainly for its own sake and emphasizes communal sharing, while the other seeks knowledge mainly for its practical utility and emphasizes private advancement. Let's look at each approach more fully.

Approach to Knowledge in Science

Centuries ago, scientists often did their work in relative isolation—sometimes not even publishing their work. Today, however, scientists usually work as part of teams, and results are frequently published in journals after review by peer scientists. This interconnectedness among researchers forms the basis for the "institution of science." Over the centuries, the institution of science has developed the following customary attitudes about truth concerning the physical world.[2]

Universalism: Science requires that all valid claims about the physical world be tested by repeatable observations and remain unaffected by the personal or social characteristics of the researcher. That is, valid claims must be objectively true regardless of social class, race, religion, and the like.

Communal ownership: Science seeks to add all discoveries to a common human heritage of knowledge for everyone. Thus, a scientist cannot control the use or communication of a discovery except to withhold part or all of it just long enough to allow the normal publication process to proceed. That way the scientist will at least gain public recognition as the discoverer. The scientific community in general should try diligently to make its findings available to everyone without barriers of cost or geographical distribution.

Disinterestedness: Science tries to seek and publish knowledge without undue influence by economic, political, or ideological considerations. Of course, an individual scientist may pursue research partly out of curiosity, benevolence, desire for recognition, and the like. However, the results of the effort need to stay untangled from those motivations.

Organized skepticism: Claims to scientific truth carry little weight unless supported by verifiable evidence. While new claims resting on little evidence need not be dismissed completely, those claims should be considered merely tentative. Conversely, even well-accepted theories sometimes have to be modified or rejected on the basis of new evidence.

These customary attitudes sound fine in theory. Does practice meet the theoretical ideal?

Recognition from Scientific Publication

Most scientists pursue their work at least partly from interior desires to learn, to create, and to share the fruits of discovery for the benefit of humankind. However, many scientists also derive their motivation from exterior rewards as well. Since communal ownership of scientific knowledge usually keeps a scientist from making much money from a discovery, the exterior reward for innovation lies mainly in public recognition. This recognition goes to those who show they were first to observe and recognize the significance of important new knowledge.

The reward for publishing first gives a healthy energy and originality to many scientific pursuits. While recognizing this fact, scientists commonly remain suspicious of efforts to win professional or public fame.[3] Sigmund Freud, who first put psychology on a sound scientific footing, described the reward for publishing first as an "unworthy and puerile" motivation for scientific effort.[4] Nevertheless, people deeply need concrete expressions of approval for what they do. Scientists are no exception. Deep down, relatively few are completely convinced of the basic worth of their work.[5] As a result, the drive to publish first can become an end in itself rather than a means to creativity. Is it really such a big deal when one scientist publishes a few days or weeks before someone else? Some observers assume these problems have become common only recently. Yet even a superficial reading of scientific history shows otherwise. Over fifty years ago, the noted sociologist of science Robert Merton put it this way[6]: "The fact is that all of those firmly placed in the pantheon of science—Newton, Descartes, Leibniz, Pascal or Huygens, Lister, Faraday, Laplace or Davy— were caught up in passionate efforts to achieve priority and to have it publicly registered."

Whereas some scientists find this competition invigorating, others prefer to avoid it by choosing relatively "unpopular" fields. Sigmund Freud and Max Planck described with nostalgia the early days of their research in neglected areas—days in which they could work out their most important ideas without the pressures of competition.[7] Unfortunately, this approach often just leads to obscurity. Several decades ago the psychiatrist Lawrence Kubie pointed out, "success or failure, whether in specific investigations or in an entire career, may be almost accidental, with chance a major factor in determining not what is discovered, but when and by whom. . . . Yet young students are not warned that their future success may be determined by forces which are outside their own creative capacity or their willingness to work hard."[8] Some scientists learn this hard truth only through painful experience. A fraction of these withdraw from the effort altogether. More commonly, however, idealism just shrivels slowly and eventually dies. Kubie posed the hard question that remains fresh today: "Are we witnessing the development of a generation of hardened, cynical, amoral, embittered, disillusioned young scientists?"

Black and Gray in Scientific Practice

Clearly the pressures of publishing first run directly counter to some of the ideals of science regarding truth. What sorts of actions offend against these ideals?

Publishing data obtained by fraud or falsification is clearly wrong. Such actions strike at the core of scientific truth. Of course, we have to distinguish between incorrect reports and fake ones. The scientific literature is full of results that turned out to be wrong due to accidental errors. Falsified data may or may not violate physical law; the important point is that the researcher makes them up. Fraudulent data come in several varieties, and again may or may not agree with the laws of nature. Here the important point is that the data are not collected by the stated method. The fraud usually involves deleting or massaging in a way that Charles Babage once called "cooking" and "trimming."[9] While cooking or trimming data without cause is clearly unethical, more difficult ethical problems arise when scientifically plausible reasons exist to throw out or rescale certain data. Sometimes instruments have obvious but intermittent problems. Other times, changes in protocol creep into the experiment, either by design or by error. Some experiments are just too time consuming or expensive to repeat. Given the length limitations imposed by some journals, it may prove impossible to explain every detail of the analysis. In the end, the scientist has to exercise prudent judgment in reporting—the literature does not benefit from avalanches of questionable data.

Plagiarism is also a clear wrong. Plagiarism involves directly copying or paraphrasing someone else's words or results without proper citation. Most people might agree you would be plagiarizing if you wrote the following without citing Z. Z. Jones:

> *From your paper of today:* "Gases like H_2 and He do not obey the RCV equations for pressure-volume behavior under extreme conditions."

From Z. Z. Jones's published paper of 1995: "Light gases like hydrogen and helium depart from the RCV expressions describing pressure-volume behavior under extreme conditions."

In your text, the sentence structure and wording match those of Jones too closely, and you are describing something that is not widely known by most scientists. On the other hand, you need not cite someone else if you are writing about something most people know and if your wording differs sufficiently from what another person has written. For example, most people might agree you would be safe if you wrote the following:

From your paper of today: "Atoms are made of a dense nucleus containing protons and neutrons surrounded by a cloud of electrons."

From Z. Z. Jones's published book of 1995: "Atoms comprise a heavy nucleus with neutrons and protons together with a surrounding haze of light electrons.

Many scientists include in their definition of plagiarism the fairly common practice of deliberately failing to cite distinct but closely related work by others. The ethical judgment of course depends heavily upon how closely related the work is.

No one knows how common plagiarism is in the published literature. However, it's interesting that accusing someone of stealing scientific ideas seems to be more common than the stealing itself![10] Accusations of plagiarism have been well known since the time of Descartes in the 1500s, who was falsely accused of stealing ideas from Harvey, Snell, and Fermat in physiology, optics, and geometry, respectively.[11] Not all such accusations are malicious. The human mind often tends to take new ideas and package them into old boxes, thereby making them look familiar.

Approach to Knowledge in Technology

As we discussed earlier, technology seeks to apply knowledge to make items of value for the society at large. Several motivations may drive the need for such items, including monetary profit or national defense. Of course, there may also be the felt need to work selflessly for the common good, that is, benevolence.

Only benevolence can coexist successfully with all four customary attitudes of science: universalism, communal ownership, disinterestedness, and organized skepticism. Motivations of profit or national defense require that at least the ideal of communal ownership be modified or abandoned, and possibly the ideal of disinterestedness as well. These changes are not necessarily bad, and can take place in full harmony with the virtues. For example, fairness may require that key elements of an invention be kept secret so that the inventor can reap some reward for creativity. Prudence may require that details of a weapon system be kept secret from an unfriendly nation.

Intellectual Property

In the commercial world where most scientists and engineers work, it's sometimes difficult to decide when useful knowledge should be kept secret or when an inventor should control that knowledge. In the United States and many other countries, balancing between truth and fairness is done by laws governing intellectual property. Before discussing intellectual property at length, let's look a little more closely at the idea of property ownership in general. First, we should note that "ownership" differs from "possession." While "possession" refers to actual control of something, "ownership" refers only to the rights to that control. Thus, if a thief steals your calculator, that person possesses your calculator even though you still own it. Given that you own something, what kinds of rights do you have? Moral and legal scholars disagree on the exact list, but most accept that you have the right to:

1. enjoy or use it yourself
2. say who else may use it and how
3. enjoy income from its use
4. give it to someone else by sale, inheritance, or gift
5. change or modify it
6. destroy it

Approaches to intellectual property vary from nation to nation. However, the United States provides the following legal protections: patents, copyrights, and trade secrets. Let's examine the main features of these three kinds of protection. (Trademarks are also a kind of intellectual property, but they do not involve the kind of knowledge that concerns us here.)

Patents: These have historically covered inventions (like machines), substances (like various chemicals) and processes (like a method for synthesis). Recent legal decisions have extended protection to certain life forms created by advanced genetic techniques. If you hold a patent, you have the right to decide who may use, produce, or sell the patented item. However, the protection extends only to the design or application of the idea, not to the theoretical basis. Thus, you cannot prohibit publication of descriptions of the idea, and cannot prevent development of the idea into still more patentable ideas. Patents are granted only after a lengthy process of examination where you must provide specific instructions for making or applying the subject of the idea, and must prove that the idea is new and not obvious to someone trained in the field. Patents provide protection for only about twenty years.

Copyrights: These cover specific ways of expressing ideas in words. Copyrights have historically covered written books and articles, but recent legal decisions have extended protection to pictures and computer software. If you hold a copyright, you can decide who may copy the specific form of your product, but you cannot control the substance of the ideas the product contains. For example, in compilations of data you can copyright only the form of the compilation, not the data themselves. If there is only one way to express the idea (as in mathematical notation), a copyright cannot be granted. Unfortunately, this distinction between form and content becomes very fuzzy for computer software, making copyright law hard to apply in this area. Unlike patents, copyrights can be granted even if the item is not new. You can get your own copyright if you can show that you arrived at the form independently of the previous copyright. Copyrights are granted under some conditions without the filing of any forms at all, although holders can pursue legal remedies more easily by following a few simple steps for formal copyright registration. Copyrights provide protection until fifty years after the death of the last surviving author.

Trade secrets: These do not carry the same formal legal weight as patents and copyrights. No filing of any kind is required. However, many legal jurisdictions recognize that patentable ideas, lists of customers, cost and pricing data, plans for new products, and the like have commercial value when kept secret, even in the absence of formal legal protection. The value of the secret depends on how much effort someone would need to discover it independently. If you have a secret and have made diligent attempts to prevent disclosure, the courts may uphold your right to exclusive use even in the event of loss (usually due to employee disloyalty or espionage). When enforceable, trade secrets provide protection forever.

There are many good reasons for a society to protect intellectual property: for example, to encourage innovation and to prevent "free-riding" on a good idea.[12] Sometimes intellectual property laws clearly work against the best interests of the public at large; in such cases, the government steps in to minimize the difficulties. Copyright law, for example, puts special restrictions on the rights of authors by permitting "fair use" copying for purposes like news reporting, scholarship, and teaching. The law also compels authors to license certain works at noncompetitive prices under some narrow circumstances. In creating these restrictions, the law seeks to limit the financial gain of authors to that which is sufficient to promote continued innovation. In another vein, various "right-to-know" laws have been enacted around the United States in response to the public need for monitoring the environmental impact of the chemical industry. Some jurisdictions require submission of detailed data about chemicals used and created to health professionals and (to a lesser extent) to employees, even if such data are trade secrets. While such laws attempt to offer some protection to manufacturers through confidentiality agreements and the like, in the end secrecy is greatly compromised for the sake of oversight in the public interest.

A REAL-LIFE CASE: Copying Music Illegally Using the Internet

For decades U.S. copyright law has protected the rights of composers and performers to the fruits of their musical creativity and skill. Under most circumstances it is illegal to electronically copy a recorded version of a musical performance. The right to distribute or sell such performances lies with the copyright holder. Of course, private tape recording of copyrighted music has been practiced for quite some time, to the dismay of musical recording companies. However, a new method for copying on a wider scale has become available with the increasing popularity of the Internet.

The copying is typically done using a digital form of data compression called MPEG Layer 3, or MP3 for short. Freely available software takes the large amount of data stored on a compact disk and compresses it into memory space nearly a factor of 30 smaller. The resulting file of a few megabytes in size is relatively easy to load and transport over the Internet. Freely available software can then decompress the file in real time, permitting respectable-quality playback over speakers at the receiver's end. Tracker software exists that permits interested people both to advertise their compressed "library" and to see what others have. Attempts to foil such schemes have proven difficult, especially with small-scale operators.

Recording companies claim to lose $300 million annually to this form of copying. They also claim that new artists get hurt because while their music may become widely distributed via MP3 technology, their compact disks suffer slow sales. Thus, the artists could lose their financial backing.

- How seriously wrong do you believe it is to employ MP3 technology for copying?
- How much attention do you think the government (which enforces copyrights) should give to the problem?

References

Baroni, Michael. "Rounding Up the Posse in a Lawless Frontier." *The New York Times*, 8 June 1997.

Chervokas, Jason. "Internet CD Copying Tests Music Industry." *The New York Times*, 6 April 1998.

"The desire for fame is the last infirmity cast off by even the wise."

CORNELIUS TACITUS (C. A.D. 55–117), *HISTORIES*, BOOK IV

Notes

1. Robert K. Merton, "The Normative Structure of Science," *The Sociology of Science* (Chicago: University of Chicago Press, 1973), 267–278.

2. Ibid.

3. Even the incomparable Isaac Newton wrote, "If I have seen farther, it is by standing on the shoulders of giants." (Written in a letter to Robert Hooke, who was challenging Newton's claim to have invented the theory of colors.) From Alexandre Koyre, "An Unpublished Letter of Robert Hooke to Isaac Newton," *Isis* 43 (December 1952):312–337, on 315.

4. Sigmund Freud, quoted by Robert K. Merton in "Behavior Patterns of Scientists," *The Sociology of Science* (Chicago: University of Chicago Press, 1973), 325–342, on 338.

5. Merton, "Behavior Patterns of Scientists," 339.

6. Ibid., 335.

7. Ibid., 333.

8. Lawrence S. Kubie, "Some Unsolved Problems of the Scientific Career," *American Scientist* 41 (1953):596–613, 42 (1954):104–112.

9. Charles Babage in an 1830 writing, quoted by Robert K. Merton in "Priorities in Scientific Discovery," *The Sociology of Science* (Chicago: University of Chicago Press, 1973), 286–324, on 310. Fortunately, the ideal of universalism provides a useful check on such behavior. If the field of study commands sufficient interest to draw other researchers, these can repeat the stated method and at least highlight results that are incorrect.

10. Merton, "Priorities in Scientific Discovery," 313.

11. Ibid.

12. For detailed arguments regarding the morality of intellectual property, see Arthur Kiflik, "Moral Foundations of Intellectual Property Rights," in *Owning Scientific and Technical Information*, Vivian Weil and John W. Snapper, eds. (New Brunswick, N.J.: Rutgers University Press, 1989), 219–240.

Problems

1. Write a page or two describing an ethical dilemma that involves some aspect of truth on a social level that you have encountered in a job you've had. (If you've been lucky enough never to have been confronted with a problem like this, describe one that a friend or relative of yours has had.) Recommend what action you think you (or your friend/relative) should have taken, and give reasons for and against that recommendation. Note: you don't have to say what was actually done in real life (unless you want to)!

2. Each case below has a question after it.

   a. List the options/suboptions available to the main character who has to make a decision, together with the event tree flowing from each option.

   b. Recommend what you think the character should do.

CASE 10.1 Withholding Procedural Steps in Scientific Publications

"So how's it going?" asked Professor Warren Clark as he walked up to his undergraduate research assistant Leah Nonlibet. "Are you almost finished with that data set?"

Leah nodded. "Uh-huh. I should be done next Monday for sure."

"That's great!" Clark exclaimed approvingly. "We'll start writing it up for publication right away. I've been wanting to get in print for the past month. This stuff we're doing with metal-silicon compounds should really put us on the map in the mineralogy community!"

"Weren't we on the map before?" inquired Leah.

"Well, yes. But Leah, Nosce te Ipsum University is not a major research powerhouse. We don't get that much respect. A lot of my colleagues in the geology department here don't even publish. With a couple of graduate students, helped out by undergraduates like yourself, I run one of the biggest programs here." Clark waved his hand toward the rest of the laboratory. "And one of the most heavily instrumented. Still, compared with people in my field at other institutions, my operation is pretty shoestring."

"So why is this new magnetic phase we found so exciting?" asked Leah.

"Something like it has been predicted by the theorists for years," replied Clark excitedly. "But no one has ever found experimental confirmation. Now we have it. It has implications for how the Earth's magnetic field is generated, since there's so much silicon and other metals down there."

"Professor Clark, I'm only an hourly worker, but I still took some of the data for this paper. Does that mean I get my name on it as a coauthor?"

"Of course!" Clark beamed. "You deserve it! The lead author will be Marcus, since as the graduate student he took most of the data. But you'll still be in on the writing if you want. And you have to approve the final version anyway, as do all the authors." Clark then paused, and continued more deliberately. "We'll have to be careful how we write this. I have several experiments planned over the coming year or so to follow up on what we've done. The ideas behind them are pretty obvious—anyone in the field would expect to see this kind of follow-up. The trouble is, once we send our paper to a journal for peer review, it's almost certain that some of my competitors who see it will jump to try those experiments right away. They have a lot more resources than I do, both people and money. They can get the work done in half the time. That will leave us out in the cold."

"What are you going to do? Can't you just hold this paper until you get everything done?" queried Leah with concern.

Clark shook his head. "No, it's too dangerous. We stumbled onto this new crystal phase by accident, and someone else could do the same. It's never worth much to be the second to publish a discovery. No, my idea is this. Remember that one key to making this phase is precise control of the cooling step? We have to do it in stages after that hot annealing, right? Well, I figure we'll just be vague about how we do that. We'll say something like, 'the material was cooled slowly over three hours to room temperature.' In fact, we have to cool one hour at 900 degrees Celsius, thirty minutes at 500 degrees, and ninety minutes at 400 degrees before quenching suddenly to room temperature. As we learned, a normal linear cooling program won't work."

Leah frowned. "Professor Clark, is that right? I mean, I thought a scientific paper was supposed to tell enough of what you did so that other people can reproduce your work."

Clark tensed. "Leah, it's not like we're lying. Our words are literally accurate. It's true they don't tell everything, but we'll fix that up when we publish those experiments from the next year or so. By doing things this way, it should slow down my competitors enough so we can get that work into print first. They can wait a year or two to hear the whole story." Clark looked at Leah gravely. "I assume you'll be agreeing to this strategy."

◆ Should Leah agree?

Case 10.2 Ignoring Outlying Data Points

"Well, things are turning out better than I thought!" exclaimed Professor Warren Clark to his undergraduate laboratory assistant Leah

Nonlibet. "I had no idea your optical absorption data would complement Marcus's magnetic data for our new crystal phase so well! Now we should be able to announce our discovery of this new metal-silicon phase in two papers going to different journals. That way we can advertise to more of the geology community."

Leah smiled modestly. "I never thought I would get my name on two papers as just an undergraduate," she chuckled.

Clark beamed. "Well, you deserve it. You worked hard. Say, let's have a look at that spectrum again." Leah showed him the paper. It plotted light absorbance versus wavelength for the new material they had created. "Yup. It's clear as day!" Clark continued. He traced along the plot with his finger. "Look at this. The absorbance is low at long wavelengths, then jumps way up here, and continues to increase slowly as we get into the ultraviolet. Classic behavior for a semiconductor!"

Leah's smile transformed into a frown as she listened. "But Professor Clark," she began with hesitation. "The absorbance doesn't stay high throughout the ultraviolet. See this point here? Down at 300 nanometers. The absorbance drops a lot."

Clark waved his hand. "Oh, that can't be. It makes no sense. Semiconductors don't act that way. There must be a mistake in the measurement, or maybe an instrument malfunction. Try it again," he observed offhandedly.

"But I did!" Leah persisted. "Don't you see? There are two data points there. I did them on separate days, and they lie within each other's error bars."

Clark grew slightly exasperated. "Leah, it can't be. Semiconductors absorb strongly at wavelengths shorter than the one corresponding to their bandgap energy. It's not a controverted point. Oceans of data are out there to support it, backed by solid theory. The predictions out there for this crystal phase say it should be a semiconductor. Plus, Marcus's data for both magnetic behavior and electrical conductivity show classic semiconducting behavior. He even gets the temperature dependence right. And every shred of your optical data except at 300 nanometers says the same thing. If you can't get reasonable numbers for 300 nanometers, we'll just publish the work with them deleted. The graph doesn't need those data to show what we want."

Leah's voice hardened. She stood up and crossed her arms. "Professor Clark, this is a new material. We can't be completely sure exactly what it is. You can't throw out these two points at 300 nanometers on the basis of any statistical analysis—you know, t-tests. And you can't throw them out just because they don't fit your theory. They have to stay in. We can't be deceptive when we publish in the open literature."

Clark's face reddened slightly as he sought to suppress his anger.

"Leah, there's nothing deceptive about dropping these points. Yes, we owe it to the scientific community to honestly report what we did. But we also owe it to them to exercise sound judgment in discriminating good data from bad. It does no one any good to publish data we know is junk, even if we label it as junk. The literature doesn't have space for junk. And about statistical analysis—that doesn't mean much here. Statistics assume random errors, which the narrow spread of your other data shows is not a big factor here. You have a systematic error. Since your glass optics start absorbing near 300 nanometers, I think it has something to do with that." Clark eyed Leah. "Plus," he added, "you're new at this game. You just learned this experiment a few weeks ago. I could tell from the way you looked just now that you didn't even know about glass absorption. I've been doing experiments since before you were born. When it comes to interpretation, I think you need to leave it to me."

Leah dug in her heels. "I did the experiment carefully," she declared. "I stand behind my data. And with due respect, I don't like how you approach publishing. It's very self-serving. First, you decide to withhold details about our preparation procedure to keep your competitors from squeezing you out. That was hard enough for me to swallow. But now you want to throw out my data. With no good reason. I won't stand for that, and as a coauthor I won't agree to it! The points stay in unless we can identify an error to justify dropping them!"

"Leah, we don't have time for that. Someone would have to check the whole optical system. You don't have the experience to do that yet, and the other graduate students are too busy already. We might have to buy new extended-range optics, which I can't afford for the sake of one stupid measurement. And it would take months for me to get around to doing it myself!"

"You can't publish lies, no matter how long it takes to find the truth!" declared Leah firmly.

◆ What should Professor Clark do?

CASE 10.3 Reporting Toxic Discharges to the Government

"How did work go today at Tripos? You haven't said much about it in a couple of weeks," Todd Cuibono asked his girlfriend Emily Laborvincet as they sat eating supper in the restaurant. He watched as her face wrinkled. "That bad, huh?"

Emily nodded. "Todd, I'm only a sophomore. And working part time this semester. But ever since the owner of Tripos had that car accident that put her in the hospital, I've just gotten slammed with responsibility . . . so many hard decisions to make."

Todd pursed his lips. "You haven't said much about them. The last one you told me about was a while back when you had trouble with that guy who wanted a little bribe in return for giving you his company's account. Actually, you never told me what you finally did."

Emily sighed unhappily. "Well, the owner wasn't in good enough shape to ask. Since I'm the only one around to do the accounting now, I had to make the decision. I just gave him what he wanted."

"Yeah, I told you that was the best thing. No one will ever find out. And Tripos still made a good profit."

Emily looked at Todd with contempt. "This stuff just doesn't bother you at all, does it?"

Todd shrugged. "Nobody likes to do it, but it's part of life. So what's on your mind today?"

Emily sighed again. She paused and eyed Todd warily. "We had a chemical spill today at Tripos."

"What spilled?"

"Benzene. A drum fell off the back of a truck just outside the building. We lost about 12 pounds."

"Anyone hurt?"

Emily shook her head. "No. We were worried about a fire, but luckily it didn't happen. Actually, most of the stuff finally just sank into the gravel."

"So what's the big deal? It's not that expensive, is it?"

Emily stared at Todd in exasperation. "Todd, you're a chemical engineer! You should know the problem. Benzene can cause cancer. And the city of Exodus has an ordinance about spills. You're supposed to report them if they're over a certain size. This is just over the limit. And there's real paranoia about spills in this city after that big spill at Acme last month. The press is licking its chops for another feeding frenzy. Todd, Tripos just had a fire a few weeks ago. The owner is still in the hospital. We only have a few dozen employees. We can't cope with a media circus right now."

"If I remember, benzene isn't a very potent carcinogen. And it's pretty volatile. You'll probably lose most of it to the air. Does anyone know?"

"Just me, the truck driver, and one other worker who was near the loading dock. Everyone else was inside. The smell was a little strong for a while, but it's not bad now," replied Emily.

"No use hanging out your dirty laundry, then," declared Todd. "There's no upside to it. No one got hurt, and no one is likely to. The stuff isn't that nasty. Tripos has to survive, and right now when it's weakened it needs to keep its image polished."

Emily glanced around to ensure no one was listening. She then glared at Todd again with disdain. "You truly don't care, do you?

Here I'm worried about poisoning people, and maybe about breaking the law, and all you can think about is image!"

Startled by Emily's vehemence, Todd sat back in his chair. "I know about poison and I know about the law," he protested. "You asked for my opinion, and my opinion is that the spill is small, the poison isn't serious, and Tripos should try to survive."

"Actually, I didn't ask for your opinion," Emily growled though her gritted teeth. "You're the one who asked how things were going. I told you, and you offered your advice for free." Todd stared at her in blank perplexity. Emily calmed herself slightly. "I know we need to survive," she continued. "Your suggestion isn't totally unreasonable. It's just that whenever I ask you about anything, you worry about two things. Looking out for yourself, and presenting a good image. You run after these things so much, nothing else seems to matter. I mean, take image. You're always the most neatly dressed guy in any class. Just impeccable. Every day. And every word you say is chosen so carefully—except sometimes around me—calculated not to offend, and not to tip your cards too much. Everyone else thinks you're so mature, so smooth. It's not healthy to obsess that much about what people think of you."

"Emily, I don't understand what you're talking about. Let's pay the bill and go," Todd replied in a tense monotone.

◆ Should Emily report the spill?

The authors thank Joseph G. Seebauer of Lubrizol Corp. for providing some of the technical background of this case.

CASE 10.4 Plagiarism

"You wanted to see me again?" asked Celia Peccavi as she glided up after class to Terence Nonliquet, the teaching assistant for Comp Sci 110. "This term paper you handed back had a note on it."

Terence drew a deep breath. "Yes, Celia, once again you're doing what you're not supposed to on assignments."

"Oh?," she exclaimed as she drew back with mock surprise, hand over her heart. "You must think I'm such a villain!"

"I don't know if you're a villain or not," replied Terence testily. "I do know that you must have copied some of this paper out of some book."

"Terence, last time you accused me of copying Jacklyn's homework," Celia responded with pretended anger. "But your evidence was weak. You only took off 15 percent, although you shouldn't have taken off any. I hope your evidence is better this time."

"My evidence was fine last time. I should have taken off more. This time, the paper reads too choppy. Some of it is written like a normal sophomore would write, and some of it is very high quality."

"So? I wrote part of it when I was tired, and part when I was rested." Celia retorted. "That's not very firm evidence. If you think I copied, you have to show what book I got it out of."

"OK. Look at this sentence," Terence contended, pointing to a spot on her paper. "It reads, 'Massively parallel processing finds use in a wide variety of practical applications, including weather forecasting.' That's word-for-word out of the text by Jenkins on reserve for this course in the library. I checked."

Celia raised her eyebrows. "That's the best you can do? It's a pretty generic sentence. Anyone could say it."

"I think the preponderance of evidence weighs against you," countered Terence.

"Well, that's fine," Celia shot back with a hint of mockery. "So you're going to take off points because I can't prove I didn't copy? Isn't that getting things backward?"

Terence's voice turned slightly preachy. "Authors have a right to their ideas and words. If you use them, you have to reference them according to accepted conventions."

"There, you go again . . . Mr. Principle!" laughed Celia. "I do so like a man with principles!"

"Celia, plagiarism is a serious matter."

"And you're waaay too serious," purred Celia, eyes twinkling. She moved a step closer. "I think I can guess why. A little trouble with your girlfriend these days? You used to talk about her in class last semester. Now . . . never!" Celia's voice grew seductive. "Anyway, she sounds like such a stick in the mud."

"Leah is none of your business," Terence snapped, recalling unhappily his recent string of arguments with her.

"You're right. She doesn't matter to us." Celia whispered. "I'll make you another deal. Just like the one I offered over Christmas. Just one date. Go out with me just once, and I'll drop this class. I promise—scout's honor!"

"Celia, we were talking about plagiarism . . ." Terence broke in.

"Oh, that," said Celia matter of factly. "I told you Terence, your evidence is weak. Last time you took off a few points on that kind of claim. I didn't tell my uncle then, but I swear I will this time. As department head, he can make your life very unpleasant." She turned to leave. "Think about it."

◆ What should Terence do?

11

FAIRNESS: PERSON-TO-PERSON

"The sum of behavior is to retain a man's own dignity
without intruding upon the liberty of others."

FRANCIS BACON (1561–1626), *Advancement of Learning*

Fairness refers to words or actions that balance what is best for everyone concerned. Failing to maintain this balance leads to inappropriate prejudice or bias in favor of one or more people. Let's begin by considering who should decide what is fair.

Conflict of Interest

Sometimes the interests of two people in a situation run directly counter to each other. Literally speaking, the term "conflict of interest" can refer to any such clash of goals. For example, a mother may need to reconcile conflicts between her children when she is distributing a limited amount of candy. However, common usage generally restricts "conflict of interest" to cases where the person deciding has a direct interest in the decision. Thus, only if the mother were splitting the candy *and* getting some herself would we consider her to have a conflict of interest.

Conflicts of interest arise all the time in professional practice. In fact, they are sometimes impossible to avoid. While these conflicts do not always poison an ethical decision, they can pose serious dangers to fairness. The risks increase as the subject of the decision becomes more important. In our example of the mother, she would be less inclined to be fair if she were distributing expensive jewelry rather than jelly beans.

In science and engineering, conflict of interest has assumed an increasingly prominent place in the news. In debates ranging from ozone depletion to tobacco smoking to silicone breast implants, scientists from opposing sides offer data and analysis that are impossible to reconcile. Unfortunately, in the eyes of the public the reliability of data and interpretation then becomes suspect. Because scientists and engineers shoulder such a large burden of public trust, they must take special care to avoid even the appearance of improper behavior.

Here are some examples of conflict of interest that commonly occur in everyday life. An academic scientist who is testing the effectiveness of an experimental drug but holds a big financial stake in the manufacturer could be tempted into "cooking and trimming" the results to favor the company's interests. A design engineer who accepts gifts from a parts supplier could be tempted to inappropriately specify that company's low-quality parts in a new tool design. Notice that the drug tester or tool designer do not necessarily yield to the temptation. Very possibly they might make perfectly objective decisions. However, there is no question that the appearance is bad.

The best antidote to conflict of interest is full disclosure. That is, truth works to support fairness. Disclosure permits all parties to decide whether the conflict of interest is too severe, and what action should be taken to reduce or eliminate it. Scientific researchers might be required to divest themselves of significant financial holdings or disqualify themselves from the work. Technologists might face restrictions or prohibitions on the gifts they can accept.[1]

Qualitative versus Quantitative Fairness

Some moral questions of fairness involve tangible, quantifiable items like sums of money or amounts of food. Other questions involve intangible, unquantifiable items like attention to relationship or public recognition. Classical moralists therefore distinguish between quantitative fairness and qualitative fairness.[2] Of course, the balancing required by fairness can be done with more precision in quantitative cases. If you break someone's window, quantitative fairness may specify quite precisely how much you must pay to fix the damage. Purely qualitative situations do not lend themselves to this kind of analysis. If instead of breaking a window you break a heart by carelessly forgetting your sweetheart's birthday, fairness may still require you to fix the damage to the relationship. There is no good way to count up what must be done, however.

Situations that mix qualitative and quantitative aspects sometimes present the worst difficulties. These cases present the uncomfortable and nearly impossible task of quantifying the unquantifiable. The results sometimes seem to defy reason. For example, liability cases come up routinely in legal proceedings involving worker injury. A plaintiff may demand monetary damages for pain and suffering. Liability attorneys know well that juries in the United States tend to award more money for bone fractures than for soft-tissue injuries such as sprains, even when the soft-tissue injuries are far more painful and crippling. Many observers speculate that graphic x-ray images of broken bones play on jury sentiments far more effectively than abstract descriptions of sprains.

Credit or Blame in Team Projects

In science and engineering, issues of fairness commonly come up in team projects. If the project succeeds, credit must be distributed. Rewards include intangibles like notoriety and influence, and tangibles like promotion and salary increases. In classroom settings, the intangibles include knowledge gained, and the tangibles include grades. Difficulties in fairness arise in several ways.

First, the tangible reward may not account for differences in performance among the group members. Consider a student laboratory, where each lab group turns in a single report. Each student gains the intangible benefit of learning in proportion to his or her effort. However, the tangible benefit of a grade is shared in common, so that the industrious and the idle get the same grade.

Second, the project may not separate into clearly defined responsibilities for each individual. Consider a student laboratory in quantitative chemical analysis. The instructor must give a grade for how accurately the group measures the concentration of analyte in the unknown sample. Getting that concentration requires many tasks that are performed by the team— weighing, sample transfer, titration, recording, and so on. Errors in any one of the tasks usually cannot be traced, but they degrade the accuracy of the final measurement. Thus, every individual is affected by the performance of the group.

Clearly, fairness requires that all members of the group contribute to the effort in proportion to the time and talent available to them. An open, honest discussion at the outset of the effort of who can best contribute what helps to avoid misunderstandings. For long projects, such discussions may need to take place regularly as new circumstances come to light. Because tangible rewards or penalties come to the group as a whole, it may prove difficult to distribute them according to who is most responsible for the success or failure of the project. However, it's quite appropriate for the group to distribute intangibles like praise or constructive criticism where these are due.

Authorship Questions

Authorship poses just one part of team effort in technical research. Yet few questions bother researchers more deeply than whose names should appear on a published paper or patent, and in what order. Often the main reward for a researcher is public recognition. Some of this recognition comes from getting a new discovery into print before anyone else. However, most published papers or patents have more than one author, reflecting the team effort usually required for research. The order in which the authors appear gives important clues about their relative contributions to the substance of the work. Practices vary widely by discipline, but commonly the first author has contributed the most to the work, or has written the first draft of the manuscript. Succeeding authors then appear in decreasing order of importance. However, in some disciplines the leader of the team (the "principal investigator" or "corresponding author") appears last.

Even if some particular convention were adopted universally, problems would still arise because there are many ways to split the labor of research. Basically, research involves conceiving an idea, performing some experiments or computations, analyzing the results, and writing up the work. However, often many people divide this labor. In U.S. universities, for example, the research team usually consists of one or two faculty who supervise students or postdoctoral associates. The faculty manage the financial aspects of the project and often conceive the idea, while the students or postdocs perform most of the experiments and analysis. Although the paper may be written by any member of the team, the principal investigators usually approve the final wording. Others may also contribute in various ways. Some people may lend equipment to the team and possibly offer training in its use. Others may provide special materials or protocols. Still others may offer minor ideas for analysis, or may suggest further avenues for research. Technicians may keep the equipment in good working order without taking any data, or may take data according to standard procedures while having little idea what they are doing. Given all these people who contribute, how can their relative contributions be recognized? Published papers provide two mechanisms: authorship for major contributions and formal acknowledgment at the end of the paper for minor ones. Differing definitions of "major" and "minor" can lead to disputes over whether a particular person should appear as a coauthor. For those with major contributions, differing opinions over "degree of majorness" can lead to disputes over order of authorship.

Because particular situations need to be decided on the basis of particular circumstances, few general rules can be given for resolving these problems. However, they become progressively easier to sort out the earlier in the effort they are discussed. Ideally, before the work even begins there should be agreement about what each person will contribute and how that will be recognized. However, reaching an agreement just before the work is written up often serves quite satisfactorily. Waiting until after the paper or patent is written invites trouble.

Fairness in Supervising

Few things corrode the atmosphere of a work environment faster than the perception that a supervisor treats some employees better than others. No doubt, the imperfect human nature of supervisors often leads to decisions that are unfair. However, even the best supervisors soon learn what parents learn when rearing several children: a one-size-fits-all approach rarely works very well. Employees have differing interests, differing ways of interacting with others, and differing levels of aptitude, luck, and need for attention. A good supervisor must adapt his or her style to match the temperament and needs of each subordinate.

So what criterion should we set up regarding fair treatment? The criterion must remain quite general to allow for the wide variety of human work experiences, but it seems foolish to operate without a yardstick of any kind. We will take fairness in supervising to mean maintaining equity in advancing the prospects of each employee. This perspective calls for a supervisor to act as a sponsor for worker development rather than as a maximizer of worker output. A good sponsor may indeed be a good maximizer, especially in the long term. However, a sponsor tends to view workers as worthwhile in their own right rather than as a means to some production goal. "Advancing the prospects" takes a long-term focus, permitting the supervisor to rotate attention and resources among employees from time to time as needs arise.

Fortunately, management styles in the United States have developed substantially since the days of Henry Ford early in the twentieth century. In order to persuade workers to accept his new ideas about specialization in manufacturing, which led to jobs with repetitive and mind-numbing tasks, Ford paid his employees top dollar. Nevertheless, he still complained that each pair of hands came with a human being attached whose demands for fulfillment collided with goals for efficiency. Today, collaborative styles of management that involve both supervisors and employees have become popular. Smoother human relations have often followed, with clear improvements in both productivity and fulfillment.

Notice, however, that "advancing the prospects" sets up a difficult standard to reach, and it does not necessarily mean "making happy." Programs to promote worker satisfaction and empowerment can fail in at least two ways. On the one hand, these programs may simply represent an attempt by management to tell workers to stop complaining and get on board. Deep down, management may still view its employees as mere servants who cannot be trusted with any real freedom or authority. These programs usually come across as superficial, and eventually collapse under the weight of empty promises. On the other hand, empowerment programs may serve as little more than unfocused exercises in feeling good, disconnected from what actually happens on the work floor. Deep down, employees may continue to resist any real change that affects them personally. Legitimate needs for streamlining procedures, improving productivity, and eliminating unneeded personnel then run into trouble.

In short, promoting fairness in the work place becomes very situation specific and does not always equate with good feelings. Many pitfalls lie hidden to trap the unwary; fairness cannot operate independently of truth and prudence—and where emotions run high, temperance and fortitude are needed.

Fairness in Contracting with Clients

Like many people in the professions of law and medicine, engineers (and to a lesser extent, scientists) sometimes offer their technical services on a contract basis. While such engineers can be self-employed or serve as consultants, they also sometimes work as employees of larger contracting firms. The professional-client relationship has many complexities, but a key issue revolves around who has most of the authority for decision-making— the professional or the client. Several models for the relationship exist that hinge on this balance.[3]

Agency: This model vests most decision-making in the client and views the professional mainly as an advocate or "hired gun." Attorneys often function this way. The problem with agency lies in the need for professionals to exercise independent judgment, unlike the foot soldier. For example, an engineer cannot always accommodate every technical demand a client may have.

Contract: This model distributes decision-making authority fairly equally between client and professional, and assumes that the contract represents an agreement entered into freely by both parties bargaining as equals. The problem here is that sometimes clients do not bargain as true equals because they lack knowledge of the field or because they have more at stake. While this inequality occurs less for engineers and their clients than for physicians and their patients, it still may undermine the fairness of the contract.

Paternalism: This model vests most of the decision-making in the professional, following the model of a parent and child. While this model recognizes the inequality in knowledge and experience between the professional and client, it tends to substitute the professional's value system for that of the client—often with very negative consequences.

Relationships between engineers (or engineering firms) and their clients depend greatly upon the technical competence of the client, ranging all the way from near-agency for engineering clients to near-paternalism for lay clients. In the latter case, some of the problems of paternalism can be reduced by having the engineer merely present information and analysis, with the client making final decisions based on his or her own value system.

Engineers who serve clients need to watch especially carefully for conflict of interest. For example, the engineer usually looks for the largest possible fee, while the client looks for the smallest. Employees of a contracting firm may need to balance the needs of a client against the needs of the employer. In turn, all of these considerations need to be balanced against the special obligation of an engineer to the needs and safety of the ultimate users of the technology—often the general public.

A REAL-LIFE CASE: **Problems with Peer Review**

The apparatus for funding and publishing research in the United States (and much of the rest of the world) is based largely upon the peer review system. For a journal article, the process works like this. An author sends a manuscript to the editor of a journal, who in turn sends the manuscript to one or more experts in the field. These experts examine the paper and return their comments together with a recommendation about whether the paper should be published as is, modified, or rejected. The editor then makes a decision and returns the comments to the author with names removed. If the recommendation is for substantial modification, several iterations of the process may be required before a final decision is reached. In the case of a grant proposal, there is usually no iteration. The grant officer simply decides whether to fund the project based on the reviews and the selection criteria of the agency.

Peer review is intended to prevent sloppy or incorrect work from making its way into the literature or being funded. However, the system contains the potential for conflict of interest. The number of people qualified to expertly review a given paper is sometimes quite small. Thus, reviewers often receive manuscripts authored by people they know personally, either as good friends or as unfriendly competitors. Unbiased judgments can become very difficult to make in such cases. For proposals, a chronic shortage of funds at granting agencies has created a situation where virtually everyone is a significant competitor. Giving a good review to somebody else's proposal decreases the chance that your own will be funded (if you have one in submission). Moreover, reviewers can often be tempted to recommend against publication or funding and then steal the ideas for themselves. Finally, there is always the temptation to stall, which slows down a competitor's rate of publication. With a lax editor it sometimes takes a year or more from time of manuscript submission to time of decision.

- Can you think of workable modifications to the current system of peer review?
- In very new or narrow fields with only a few researchers, how should nearly unavoidable conflicts of interest be handled?

"I count him braver who overcomes his desires than him who conquers his enemies; for the hardest victory is the victory over self."

ARISTOTLE (384–322 B.C.), QUOTED IN STOBAEUS, *FLORITEGIUM*

Notes

1. For a more detailed discussion of morality and gift giving, see Robert Almeder, "Morality and Gift-Giving," in *Ethical Issues in Engineering*, Deborah G. Johnson, ed. (Englewood Cliffs, N.J.: Prentice-Hall, 1991), 327–329.

2. This distinction is often referred to by the slightly less precise terms "quantitative justice" and "qualitative justice."

3. For a more detailed discussion of these distinctions, see Michael D. Bayles, "Obligations Between Professionals and Clients," in *Ethical Issues in Engineering*, Deborah G. Johnson, ed. (Englewood Cliffs, N.J.: Prentice-Hall, 1991), 305–316.

Problems

1. Write a page or two describing an ethical dilemma involving fairness you have encountered in a job you've had. (If you've been lucky enough never to have been confronted with a problem like this, describe one that a friend or relative of yours has had.) Recommend what action you think you (or your friend/relative) should have taken, and give reasons for and against that recommendation. Note: you don't have to say what was actually done in real life (unless you want to)!

2. Each case below has a question after it.

   a. List the options/suboptions available to the main character who has to make a decision, together with the event tree flowing from each option.

   b. Recommend what you think the character should do.

CASE 11.1 Authorship

"Warren, I still say I should be first author on both papers," declared Marcus Sloane combatively to Professor Warren Clark, his thesis advisor. "Leah should definitely appear, but as second author."

Leah sat rigidly in the chair next to Marcus's in front of Clark's desk. "And I still say I should be first author on the paper with the optical data I took," she countered. "He can be first author on the one with his magnetic data. But the optical data we'll show are all mine."

"But Leah, didn't Marcus build the optical apparatus? And test it out? I thought he took a whole set of preliminary data . . ." observed Clark.

"I sure did!" Marcus burst in. "My data were crude, but they showed all the main trends. A few months ago I trained Leah to do the experiment, and she just redid the experiments with the precision we needed for publication. She was basically a technician."

"I was not!" Leah shot back. "I know I'm only an undergraduate hourly worker, but I made real contributions, too! I was the one who perfected the alignment procedure, and I was the one who figured out we needed an extra redpass filter for long-wavelength measurements. I also wrote the computer software for the data analysis." She turned to face Marcus. "Anyway, my technique is a lot more careful than yours," she sneered. "Your data never look as good as mine."

Clark intervened before Marcus exploded at this jab. "OK, OK. Let's avoid insults," he scolded. "That's not going to get us anywhere. Now, let me ask—who wrote the first draft of the paper?"

Marcus and Leah glanced uncomfortably at each other.

"Actually, we both did," responded Leah. "I wrote part of the introduction, the whole apparatus and procedure section, and part of the results. Marcus wrote the rest . . . the remaining results, the discussion, and conclusion."

"Look, Warren," Marcus interjected earnestly. "I got this project started two years ago. I put together the main part of the apparatus. And I discovered this new crystalline phase. She just came in six months ago as an undergraduate and did mostly what I told her. Now we're going to publish two really nice papers. Am I supposed to split the first authorship position with her? It will look like she did half the work! That's not fair to me. And I need these papers for my resume, since I'm looking for a job next fall."

"The paper with your name on first is bigger, Marcus, and it's going to a more prestigious journal," Clark observed.

"But that's not enough," Marcus protested. "I did way, way more work than Leah. And you know how it goes . . . not everyone knows a great journal from a good one. And hardly anyone looks at paper length. Recruiters are just going to see that I'm first author on one paper, and second on another. That's misleading."

"But it's also misleading to put your name first on a paper that uses important procedures I developed, shows absolutely no data that you actually took, and was half written by me," retorted Leah.

Clark leaned back in his chair with his hands behind his head. He greatly disliked conflict. "Well, at least no one is questioning my position as last author," he chuckled with faint irony. "In this field, the one who conceives the idea and pays the bills always shows up last." Then he sat forward. "How about a simple solution? Let's flip a coin," he offered a bit timidly. The contorted faces that greeted this suggestion dashed his hope for a clean solution. "Well," he continued. "I'm the principal investigator, so I guess ultimately I need to make the decision. No matter what I do, someone is going to be unhappy. I'll think about it."

♦ What should Professor Clark do?

CASE 11.2 The Common Good: Not in My Backyard

Myra Weltschmerz knitted her brow as she drove up to Sabra Malafide's house after finishing another day of classes at Penseroso University. She could not understand why her friend Dolores Sola was standing motionlessly in front of the mulch pile behind the duplex next door. Myra switched off the engine and quickly got out of the car.

"I know it's warm for this time of year, Dolores," she called with hesitation. "But it's sort of breezy. You don't look good. Is your ex-husband getting you down again?"

"He always gets me down" Dolores responded. "But that's not the problem right now. It's this smelly pile."

"I never paid attention to this," said Myra. "Is that really what it is? What's it doing here?"

"The owner of this duplex—it's his idea of helping out the neighbors while making a little money on the side" replied Dolores. "You know how the city picks up grass clippings and other yard waste only once per year? And how burn piles are illegal? Well, about a dozen of the neighbors in these nearby houses decided they weren't too fond of mulching all the stuff from their yards. It's a little unsightly, and you have to know how to do it. Also, some of the neighbors are getting up in years and want to have someone else take care of it. So the owner here offered to take the yard waste off their hands for a modest fee and mulch it himself. It's actually a pretty good idea. The neighbors think it's great, and the owner makes a little extra cash from part of his land that isn't really being used anyway."

"So what's the problem?" inquired Myra.

"Well, sometimes the pile smells. Usually not horrible. But if the wind blows a certain way—like it is today—the stink makes life unbearable for me and my children." She pointed to the basement of Mrs. Malafide's house. "I wanted to air out my apartment today, because it's so warm outside. But I got the smell again, just like last summer."

"I thought mulch piles weren't supposed to smell. Have you talked to the guy?"

"Uh-huh, and he agrees that well-maintained mulch piles shouldn't smell. The trouble is that he owns several complexes. He's very busy, and can't always get the lime added right when the neighbors drop off their stuff. He promises to get the job done, but somehow I still wind up with a smelly apartment sometimes. My kids really hate it and get hard to handle."

"Can't you just close the window?" asked Myra.

"Today, yes," sighed Dolores. "But not during the summer. We just suffocate. And my ex is forever behind with child support, so I can't afford to turn on the air conditioner."

Myra became indignant. "I thought there were laws against deadbeats. Anyway, can't you talk with Mrs. Malafide? She lives upstairs from you, so she must smell the same thing. And she owns your place, so she should be fighting for you."

"I tried. It turns out her nose doesn't work very well. Like a lot of old people. So she doesn't smell much. Plus, she runs her air conditioner in the summer. And she sends mulch over there herself sometimes. So she won't do anything."

"Why don't you put more pressure on the duplex owner, like sue him or something?"

Dolores paused for a moment in thought. Then she looked up and said, "It's not as simple as you think. I've been pursuing my ex to pay up endlessly, but the legal system just doesn't work very efficiently. As for suing the owner, I don't know if I have that kind of energy or money. Anyway, he *is* trying to do something nice for the community, and I think he really does mean well. He just can't do everything. He could stop offering the mulching service, and that makes life harder for the neighbors. My lease with Mrs. Malafide is up in a few months. It might be easier for me just to move somewhere else with the kids."

◆ What should Dolores do?

CASE 11.3 Distributing Financial Responsibility after Accidents

"You know, I always get nervous in these fancy curio shops," remarked Martin Diesirae to Emily Laborvincet as they made their way through the narrow aisle. "And that's why," he continued as he pointed toward the end of the aisle at a sign reading "You Break, You Buy."

Emily nodded. "Thanks for coming," she replied. "I want to make sure I get something nice for my sister's birthday, and it's always hard to decide with all this selection. A second opinion helps. And Todd refuses to go shopping for things like this unless I beg him."

"So what's the deal between you and him?" Martin inquired. "Are you still together? I mean, we used to just have dinner together. Now we're shopping, too. Eventually something has to give."

"I don't know, Martin," she replied as she picked up a glass figurine. "He's getting on my nerves more and more. But we haven't split yet. Not like you and Myra, anyway." She put the figurine back down in favor of another. "Let's not talk about it now. What do you think of this one?" she asked, holding it up.

Martin shrugged. "OK, I guess."

"Come on, Martin!" chided Emily. "You hardly even looked. Here . . . isn't it nice?"

Martin took the figurine and examined it as Emily picked up another. He quickly grew bored, and changed the subject. "You were lucky you had classes today and didn't have to come to work. Things were really hopping at Tripos. It's getting to be like that all the time."

Emily nodded, putting down the figure she held. "Yeah, it gets me pretty stressed out, too."

Lost in his train of thought and looking blankly at the display in front of him, Martin mechanically offered the figurine he held back to Emily. "Yeah, I've got to find some way to get that moving again. . . ." As he felt Emily grasp the piece, he let go. Instantly it slipped from her fingers and shattered on the floor. Both of them gaped in horror at the remnants.

"Now look what you did!" Emily moaned. Reflexively she stepped back slightly to lift her foot for a better look. As she did so, she bumped a crystal clock sitting near the edge of another shelf. It also fell to the ground and shattered.

Both of them stared in stunned silence. A saleswoman appeared immediately. "What happened?" she asked somewhat harshly.

"I . . . we dropped this figurine," stammered Emily, "and I thought a piece hit my ankle. But the aisle was too narrow. As I checked— my ankle—the clock . . . it fell."

The saleswoman crossed her arms and narrowed her eyes. "You have to be more careful," she scolded. "There must be a full moon. That's the fourth breakage in two days."

Martin pointed at the sign over the aisle. "Does that mean we have to pay?" he ventured apprehensively.

The saleswoman nodded. "I'm afraid so. I'm sorry." She stooped and found the price stickers in the remnants. "Let's see. The figurine was sixteen dollars, and the clock . . . was seventy eight. Plus tax." She stood up. "I'll go get a broom. We'll settle after I clear the shards. They're too dangerous to leave here." She hurried off.

"I can't believe what you did!" burst out Martin. "Dropping that figurine was bad enough . . . but the clock! With tax this will be a hundred dollars!"

"What *I* did! Martin! You weren't even looking when you gave that thing to me. I didn't have it yet. You just let go! That's not my fault!"

Martin's voice hardened. "I was sure you had it. I don't know why you couldn't hold on. I think the bill is yours."

"No way!" she shot back. "You didn't give me time to get a grip. You've got to pay for at least part of the figurine. And you didn't even ask me how I was! I could have been hit by glass! Fine companion you are!"

"Were you hit?" he asked impatiently.

She stooped to examine her legs carefully. "No," she snapped, standing up. "But I could have been."

"Well, I didn't make you back into that clock," Martin huffed in anger. "I hope you don't want something for that too."

◆ Should Emily and Martin split the bill? How?

CASE 11.4 Paying for a Shared Meal

"Hey Terence, when are you going to get that bill split up?" came the yell from across the long table at Iona Chin's Chinese Restaurant. Laughter rippled through the group of twelve friends seated around the table.

"I'm working on it—just give me time!" Terence Nonliquet scolded jokingly. "You'll make my job easier if one of you just volunteers to pay the whole thing yourself!" More good-natured laughter followed, and the friends returned to their noisy conversation. "It's always a pain to split these bills up," he muttered to Leah Nonlibet, his girlfriend sitting next to him. "And it's hard to think with all this noise. Let's see if I have this straight. Eight people were drinking from the pitchers of soda, three ordered individual beers, and one person had water. Everyone ate off the common plates of food, though."

"Except me," Leah broke in quietly. "I hardly ate anything."

"Yeah, I noticed," responded Terence. "What's wrong? Not feeling well?"

"No, the food was just too spicy for me. I thought some of the dishes would be mild, but it seems like everything burned my tongue."

"The chef here is a little free with the hot stuff," Terence agreed. "It's too bad, though. I thought you would like the place." He started to scribble calculations on a napkin.

Leah sat quietly for a moment, thinking. Then she leaned over to Terence. "So, what kind of contribution are you going to want from me for this? You know, for the two of us?"

Terence's brow furrowed as he continued writing. Then he paused and looked at her with irritation. "Are we going to start that all over again?"

Leah pulled back with indignation. "What do you mean, 'that'? It was a simple enough question!" she rasped hoarsely, trying to avoid attention from the others at the table.

"Leah, we went through this a couple of weeks ago, remember? When I asked you for some of the money you made in Professor Clark's lab working on that special project? I helped you a lot, and asked you for a fair share of what you made. Then we had a big debate about whether that was proper. I thought the deal we cut was that I'd drop my request for the money, but you'd start contributing more when we go out to eat and stuff."

"That's what I'm trying to do!" contended Leah. "I was just asking you what you want! We didn't agree on exactly how much more."

"Leah, we both ate the same food and both had soda. Isn't it obvious we should both put in the same amount?"

"I'm not a mind reader. I was just making sure I understood what you wanted. And it's not so obvious to me. I hardly ate anything. I don't think it's fair that I should pay full freight when I didn't eat. I'm still hungry!"

"You can't do that!" Terence scolded. "What am I supposed to do? Get out a scale and weigh everyone's portion before they eat it? Or charge more if they have seconds? It's not practical!"

"I'm not talking about a scale. But people who ordered their own beers are going to pay for those, right?"

"Yeah, but I'm not going to ask how much each of the soda drinkers swilled out of the pitcher!" Terence exclaimed. "Again, it's not practical."

"Terence, it's not impractical to see that I ate almost nothing. I tried a tiny spoon of each thing I thought was mild. The sum total doesn't fill even a quarter of my plate!"

"Leah, everyone is having a good time, except maybe you." He nodded in the direction of the others, who were still chattering away. "I'm not going to interrupt their fun to ask about whether you should pay or not. What if people don't agree? It spoils the atmosphere."

Leah frowned. "I wouldn't worry about what they think. They'll never know if you leave me out. I'm only one person out of twelve. That's less than a 10-percent change in each person's bill. No one will ever catch it."

Terence pondered for a moment. "There's got to be a better way to handle these money things with Leah," he muttered to himself under his breath.

◆ What should Terence do?

12

FAIRNESS: SOCIAL

"In the first place [science] cannot conflict with ethics.... Man, then, cannot be happy through science, but today he can much less be happy without it."

HENRI POINCARÉ (1854–1912), *VALUE OF SCIENCE*

At the level of society in general, science and engineering present less of a philosophical split for fairness than they do for truth. Nevertheless, fairness poses some very difficult issues. We will discuss several that concern knowledge and its ownership, care of the environment, and the professional advice of technical experts.

Intellectual Property and the Society

The ability to own intellectual property provides a financial incentive for creating new things. This incentive has a significant weakness, however, because it promotes innovations mainly in areas where money can most likely be made. Lines of research that benefit either very small groups or large groups having no money tend to be neglected. Drug development for very rare diseases serves as a prime example. Regardless of how much the victims of such diseases would benefit, the market often is too small to justify the effort on a financial basis alone. Similar problems have plagued the development of high-efficiency cooking stoves for poverty-stricken countries.[1] Impoverished families simply cannot afford to buy better products, no matter how much deforestation and other environmental degradation would be prevented in the long term. Where needs like these become severe enough, governments sometimes need to step in to encourage that needed research happens. Incentives include stronger monopoly rights or direct financial payments or credits. Of course, these arrangements require a case-by-case evaluation that balances the needs of the targeted group with the needs of the society at large (which is affected by the subsidies the government offers or the rights it gives away).

Related issues arise when a company develops a patentable idea but then decides not to pursue production. Sometimes the company will obtain a patent on the idea anyway simply to prevent competitors from using it (or to generate licensing fees in the future). Such a strategy is called "strategic patenting." The morality of this practice depends on what the idea is. Strategic patents on innovations that many people might need seem inappropriate. Fortunately, however, great needs usually translate into lucrative markets. Because of this fact, the government has not felt strong pressure to step in and regulate the practice. Hence, provisions for "fair use" for scholarly and other limited purposes do not exist for patent law as they do for copyright law.

Fairness in handling intellectual property becomes especially difficult in university research.[2] Competition for government research funds has become extremely keen. Thus, financially squeezed universities have turned increasingly to private companies for funding. Although companies sometimes donate money for philanthropy, more commonly they view donations as investments in research that should generate a monetary return. Thus, contracts are written covering who will own the intellectual property and the licensing rights. When two companies enter into a contract like this, the ethical issues usually remain straightforward because there are only two parties, and both operate for profit. Much greater complexity arises when one contractor is a university, for two reasons. First, more parties are involved on the university side, each having its own interests. Graduate students and postdoctoral associates want education. Faculty researchers want to publish their discoveries before competitors and possibly to collect a fraction of patent royalties. Second, the university wants to preserve an atmosphere that is good for education and scholarship while keeping finances in order.

On the level of the individual contract, corporately sponsored university research is governed by complex, individually tailored agreements over publication rights, patent rights, licensing rights, and the like to ensure fairness. However, broader societal issues of fairness relating to the pursuit of knowledge can remain unresolved. We pointed out earlier how the chance to obtain intellectual property can lure companies into areas where money can be made. Individual faculty are not immune from these pressures. The neglect of less lucrative areas of research can pose serious long-term problems for society; this issue has begun to receive attention only recently.

Environmental Issues

Certain features of our environment are held in common, including air, water, space, and appearance. Environmental degradation can include destruction of existing resources, such as loss of biodiversity through destruction of jungles, agricultural land through urbanization, and aesthetics through clear-cutting of forests. Degradation can also include introduction of new pollutants—not only chemicals but also noise and light (that obscures the stars at night). Ethical issues of fairness therefore enter in unavoidably, since degradation by a few lowers the quality of life for the many.

It is commonly supposed that serious environmental concerns have arisen only in the past few decades. Indeed, the federal government in the United States began to regulate environmental issues seriously only in the late 1960s. The "energy crisis" of the early 1970s, the numerous toxic waste disasters of the late 1970s, the Three Mile Island nuclear power plant incident of the 1980s, and the Antarctic "ozone hole" of the 1990s have heightened public awareness of how vulnerable the environment is like never before. Nevertheless, people have contended with environmental degradation for centuries, particularly near fragile habitats like deserts. Overuse of land for agriculture undoubtedly contributed to the decline of advanced Indian culture in the southwestern United States five to six hundred years ago, and of certain peoples on the Arabian peninsula even earlier. The Industrial Revolution of the 1800s in western Europe caused unprecedented overcrowding and severe air pollution.

Yet it remains true that continual increases in the Earth's population, average human life span, and the technological capacity for degradation have magnified potential environmental problems to a truly global scale. Widespread deforestation, ozone depletion, and global warming are only a few examples. Such issues have ethical dimensions that are extremely complex; the brief space we have here can highlight only a few of them.

For example, let's consider what it means for the environment to be "clean" of harmful substances. For a given substance, we first need to decide at what level pollution begins to have significant moral importance. Depending on the hierarchy of values we choose, we can define several thresholds. The threshold could occur at the level at which:

1. the pollutant becomes detectable by the best methods
2. the pollutant is present in nature
3. the pollutant begins to pose significant risk for human health
4. the risk for harm exceeds that for other risks people commonly accept
5. the cost of eliminating the pollutant becomes prohibitive

These thresholds have been loosely ranked in decreasing order of strictness. The first two represent concrete numbers that include no consideration of probability or risk and require no balancing against other considerations. However, the third and fourth thresholds do include probability and implicitly balance health against other things. The balancing enters in through the words "significant" and "commonly accept," where someone must weigh the likelihood of harm against factors like inconvenience and cost. The fifth threshold makes this balancing against financial cost explicit, and is listed separately mainly because of its common usage.

Not surprisingly, the last three thresholds tend to generate far more controversy than the first two by requiring a balancing process that different people do in different ways. Interestingly, the first two thresholds seem to avoid this problem, but in fact they do not. Their relative strictness usually requires considerable effort and financial cost to satisfy. Since all societies have limited resources, the time and money devoted to environmental preservation cannot be used for other purposes—to feed the hungry, for example.

Experts and Paternalism

Chapter 11 has already examined the idea of paternalism in business contracts. In a paternalistic contract between a client with little technical expertise and a professional advisor, most decision-making is done by the professional, following the model of a parent and child. Unfortunately, a paternalistic contract tends to substitute the professional's value system for that of the client—often with very negative consequences. Actually, this problem has a much broader social dimension. We have often said in this book that the increasing technical complexity and specialization of our society place ever-greater burdens of public trust on scientists and engineers. Indeed, Western culture is rapidly transforming into a culture of experts, where key policy decisions are made and explained by highly trained specialists. Although this trend often squares with the demands of prudence, dangers lurk under the surface. Experts can easily fall into a mindset that assumes unlimited knowledge and wisdom and look down on those who are not "in the know."

Against this kind of arrogance one wit has cynically retorted, "An expert is someone who carefully avoids minor errors while sweeping on to the grand fallacy." While such sarcasm undoubtedly goes too far, the history of science is full of reports by reputable researchers that turned out to be nonsense, for example, n-rays, polywater, cold fusion, and solute "imprinting" of water at zero concentration.[3] More common and less obviously wrong have been the passing fads in research that started with discoveries that were supposed to bring grand new opportunities for human advancement. Such claims have been encouraged in the United States by a funding system that often values newness over depth. Unfortunately, technology is no more immune to these pressures than is science. Continued growth of the nuclear power industry and widespread replacement in integrated circuits of silicon by gallium arsenide were once trumpeted as waves of a future that never arrived.

Compounding these problems is the fact that scientists and engineers work in a society that sometimes seems superstitious or paranoid. Recent controversies over "alternative medicine" and the health effects of low-frequency electromagnetic fields (EMFs) offer just two examples of debates where logic seems to take a holiday. In response, scientists and engineers can be tempted to think, "we know what we're talking about and you obviously don't, so you should just keep quiet and listen to us." Such paternalism offends against fairness by not giving others the respect due to them, and stands in the way of what the common good really requires—clearer and more patient explanation of facts and physical principles. This attitude can also lead to the pursuit of lines of research that are out of touch with human needs.

Social Aspects of Employment

Although some scientists and engineers run sole proprietorships or partnerships, most work for corporations as employees. We discussed some of the person-to-person aspects of the employer-employee relationship in Chapter 11. However, this relationship has important social aspects as well. Here we will touch on just a few of them.

One concerns the long-term relationship between employers and their employees. Many workers, particularly those with highly developed skills, view their jobs as more than merely sources of income. They look to their work as a source of personal fulfillment as well. This fulfillment may have many origins, but chief among them is often the successful completion of tasks as part of a team. For such efforts to work well, employees must develop a sense of trust and loyalty toward the team and the team toward them. Both trust and loyalty draw on all the classical virtues, but loyalty draws especially heavily on fairness and fortitude. Fairness requires that a bond forged with careful effort over time not be cast aside lightly, while fortitude provides the means for sustaining that effort in the face of difficulty. Many observers have criticized the recent trend toward a loosening of those bonds, as shown by the increasing tendency of corporations to lay off "excess" employees as well as by the increasing rate of "job-hopping" by workers. Few people question the acceptability of these practices if the reasons are sufficiently good. However, employment statistics as well as a great deal of anecdotal evidence suggest that more and more often, the reasons for layoffs or job-hopping are not sufficiently good. This phenomenon comes from a widespread rootlessness and lack of commitment, and leads to a decay in the social bonds that keep a society from flying apart.

A second aspect of employer-employee relations concerns the increasing tension between the demands of work and personal life. Statistics and anecdotal evidence over recent years suggest that average worker satisfaction has declined and stress has risen. Social scientists are still trying to quantify exactly how bad the tensions are and where they come from. No doubt the rise of two-working-parent families, increased commuting times, and longer hours demanded by a more competitive work environment all play a role. As bearers of responsibility for a great deal of research, design, production, and management, scientists and engineers seem particularly vulnerable to these stresses. Fairness helps keep the balance between the legitimate production needs of employers and the legitimate personal aspirations of employees, and has led to flexible working hours, work-at-home arrangements, job sharing, and company-provided daycare. Nevertheless, some companies and lines of work offer more such opportunities than others. Moreover, in the end a day still contains only twenty-four hours into which people can cram only so much activity.

A third aspect of employer-employee relations concerns the continuing underrepresentation of women and certain racial or ethnic groups in some segments of the work place. Science and engineering have proven unusually resistant to penetration by these groups. This issue has led to a continuing and often nasty debate that we will not reproduce here. Clearly fairness requires that opportunities for employment and advancement should be equal regardless of gender, race, or ethnic background. The more difficult questions revolve around how equal opportunity can be ensured and whether equal opportunity necessarily implies equal results. Other questions focus on the injustices that created inequalities, who perpetrated them, whether restitution is due, and over what length of time.

A REAL-LIFE CASE: Environmental Cleanup— Problems with the Superfund

In response to several notorious incidents of toxic waste dumping during the 1970s, the U.S. Congress enacted in 1980 the Comprehensive Environmental Response, Compensation, and Liability Act— known more commonly as the Superfund Act. The law was intended to begin rapid cleanup action at hazardous sites while forcing polluters to foot the bill. To accomplish this purpose, the law created a "superfund" financed by a tax on the chemical industry. This fund served two functions: to bankroll early stages of site cleanup while those responsible were being sued for compensation, and to finance cleanup of "orphan" sites where no guilty party could be found.

Unfortunately, most observers agree that the Superfund Act has not served its intended purpose. Faced with huge cleanup costs, many companies accused of bearing responsibility first deny it, and then sue their insurance company. The legal proceedings become extraordinarily drawn out. To compound the problem, the Superfund Act employed a "joint and several liability" approach, meaning that any party to the dumping, no matter how insignificant, could be held legally responsible for the entire cost of the cleanup. In the face of such potentially crippling penalties, joint contributors to pollution at a given site often engage in lengthy legal battles with each other. Finally, orphan sites have often lain untouched because most of the superfund's resources have been consumed by legal costs.

Although there is widespread recognition that the law is not working well (except for attorneys), there is little agreement about how to fix it. The American landscape is still marred by many untouched waste dumps. Congress is continuing to look at the problem, but a solution does not appear to be coming soon.

◆ What do you think is the most important problem with the Superfund Act?

◆ How would you fix the law?

References

Koshland, D. E. "Toxic Chemicals and Toxic Laws." *Science* 253 (1991):949.

Voorst, B. V. "Toxic Dumps: The Lawyers' Money Pit." *Time*, 13 September 1993, 63–64.

"I want no money raised by injustice."

Canute the Great, king of England and Denmark (995–1035), "Letter of State"

Notes

1. Amulya K. N. Reddy and Jose Goldemberg, "Energy for the Developing World," *Scientific American*, September 1990, 111.

2. A different situation than described by Vannevar Bush, director of the U.S. Office of Scientific Research and Development, in the report *Science: The Endless Frontier* (Washington, D.C.: U.S. Government Printing Office, 1945).

3. See *Pathological Science* by Irving Langmuir (Schenectady, N.Y.: General Electric Co., 1968).

Problems

1. Write a page or two describing an ethical dilemma that involves some aspect of fairness on a social level that you have encountered in a job you've had. (If you've been lucky enough never to have been confronted with a problem like this, describe one that a friend or relative of yours has had.) Recommend what action you think you (or your friend/relative) should have taken, and give reasons for and

against that recommendation. Note: you don't have to say what was actually done in real life (unless you want to)!

2. Each case below has a question after it.

   a. List the options/suboptions available to the main character who has to make a decision, together with the event tree flowing from each option.

   b. Recommend what you think the character should do.

CASE 12.1 Deciding the Layoff Target

Although the workday had ended, Martin Diesirae and Emily Laborvincet found themselves still sitting in the manager's office of Tripos Metal Polish, trying to finish the day's tasks. Suddenly, in frustration Martin disgustedly threw his pencil down on the desk in front of him. "Whew!" he exclaimed. "My brain is dead. I don't want to do anything else today."

"Mmm," grunted Emily faintly as she continued to tap on her computer keyboard.

Martin sat pondering for a moment. He turned for a few moments to watch Emily, who had her back to him. "You know, Emily," he ventured with hesitation, "it's been a couple of weeks since the fire here. You and I have just gone about our business, trying to get things back in shape. But . . . well . . . I wanted to apologize. You know, during the fire, for trying to drag you out of here when you were trying to save the company files."

Emily turned to face him, peering intently into his face. "Oh, I guess it wasn't totally your fault. You were scared that I wouldn't make it out."

"Yeah," Martin agreed. Then he rubbed his jaw. "But I didn't count on your right hook. It hit me like a sack of bricks! With dead aim! Where did you get it?"

"My brothers taught me," Emily laughed. "It doesn't still hurt, does it?"

Martin shook his head. "Nah. Only for a day or two. I'm sure glad you saved the files and got out! You're fearless!"

Emily raised her eyebrows. "Hardly. Just after the fire, Rolf—from production—gave me quite a scare."

"Rolf? The big guy? What did he do?"

Emily stopped short and glanced away nervously. "It doesn't matter," she mumbled. "It's over now."

Martin paused for a moment, puzzled. Then he sat up with renewed enthusiasm. "Hey, Emily, I'm hungry. Let's go to that Chinese place across the street again." Emily continued to look away, and said nothing. "Well?" Martin inquired.

"Martin, I can't. I'm sorry." Emily responded, glancing at him and sitting back in her chair.

"Can't? What's that supposed to mean?" Martin demanded.

"I've decided I can't go out with you any more. I finally figured it out after you refused to pay for any of the stuff we dropped by accident in the curio shop. You're a nice guy, and I like you. I want to be friends. I just don't want to date anymore. You . . ."

"Ah! The old 'just friends' line," Martin Diesirae interrupted with a hint of sarcasm. "The standard procedure for 'politely' dumping some poor guy."

"Don't be like that Martin," chided Emily. "Technically there can't be a 'dump' because technically we were never an item. I still have a boyfriend, remember."

"So that's really it . . . you've decided to stick it out with him."

"No, I haven't," Emily contended. "I told you, I still don't know where that's going. But that's not the issue here. You just have too many rough edges for me. Your temper is too quick. Your opinion of yourself is a little too high. And in your heart, I'm not sure you respect *any* woman very much, not just me. When I add it all up, I'm looking for a different kind of man."

A long silence ensued. Finally Martin sat up in his chair and observed, "OK. I see now. There's no use talking about it anymore." He glanced up at the office clock. "It's late. We should be going home. You said earlier you had some other business to talk about. Let's get it over with."

"Martin, I'm serious when I say I still want to be friends," she entreated.

"We'll be friends, don't worry," he replied stiffly. "Let's start by getting today's business over with."

"OK, if that's what you want," Emily responded reluctantly. "We have a big problem. I already talked it over with Monica as much as I could with her lying in the hospital. I do all the balance sheets and personnel stuff, but she owns this business, so I needed to get her OK. Tripos has a cash flow problem. That fire hit our plant a couple of weeks ago now, and knocked out about half our production capacity, as you know. We haven't been able to fill a lot of orders, but we're still paying all our employees. We can't do it anymore. There's not enough cash, if we want to pay all the contractors to get our plant back in shape."

"Can't we get a loan?" asked Martin.

"I tried, with as much help as I could get from Monica. We can get some, but not enough. The banks are worried about the ability of Tripos to get back on its feet. After all, both our owner and our chief sales manager are laid up in the hospital for months from a car accident, and then half of our only plant burns up. We're not a good risk."

"So, what now?"

"We employ fifty-one people now," continued Emily. "I did the numbers. Roughly ten will have to go on temporary layoff within two weeks. If things don't improve, we'll have to lose another ten a month after that. Until our plant is fit for full production again."

"So who gets the ax? And who decides?" asked Martin.

"Ultimately Monica decides. But she hasn't been here for a while, and her mind is still too fuzzy to deal with the details of balance sheets. Remember, she got pneumonia again last week. So she asked me to consult with some people—mainly you, as temporary sales manager; Conrad, as head lab chemist; and Selma, as head of production. Then I'll prepare a plan and present it to Monica. I think she'll approve whatever I recommend."

Martin's face showed concern. "How are you going to decide? Seniority? We have some young people who are really good, and could leave the company if they're laid off. And some of our newer workers are living hand-to-mouth, with kids and mortgages. Layoff would be a disaster for them. Maybe it's just fairest to drop the people who work on the lines that are knocked out."

"Yeah, Martin, but like you said—some of those have seniority and are really good. Actually, probably most of the people in Conrad's chemistry lab should go. We don't need new product development right now. Maybe just a quality control chemist."

"You mean lay off Conrad before one of his junior chemists? That won't sit well! What if he goes? He's good enough to get a job anywhere! We have to think long term, not just short term."

♦ What should Emily recommend?

Case 12.2 Publishing in Widely Accessed Journals

"I reviewed this paper like you asked," said Leah Nonlibet as she placed a packet on Professor Warren Clark's desk. "I worked through the analytical method the authors outline. I can't find any mistakes. Actually, it works pretty well once you wade through the bad English to figure out what they're saying."

"Thanks for helping me with the review," responded Clark. "Sometimes I get a whole bunch of journal articles to review at once. I just can't keep up without help. And I didn't feel I could ask Marcus to help. He's busy trying to get data for his thesis so he can graduate in December. And he's still fuming that I made you first author over him on that one paper."

Leah nodded. "Yeah. He'll hardly talk to me these days. I didn't know it would be such a big deal for him."

"It was a big deal for you," Clark pointed out quickly. "Enough that you were willing to become a little more accommodating when it came to dropping those data points you were defending with such passion."

Leah flashed a faint smile. "Well, I still don't like the idea. But the whole situation was just getting too messy, and I could tell you needed help to solve it. I'm glad it's behind us now."

"Me too," Clark agreed. "Now I have a new problem, although thankfully smaller. It's about this paper you just reviewed. The authors are from Antipodea. The country is poor, and they have hardly any resources. As you saw, the whole paper just outlines a method for analyzing data from optical measurements on minerals. The method is intended for a hand calculator. Leah, I would never use this. Neither would any colleague I rub shoulders with. We all have personal computers to do this kind of thing, in about 1 percent of the time."

"Yeah, but they don't have many personal computers in Antipodea, or in a lot of other countries. So the method could be useful there," Leah observed.

"I know. I wanted you to check whether the method was valid first," said Clark. "If it was invalid, the review would be simple to write. But now I have to make a tough choice."

"What's so hard? The method is OK. You can just recommend its publication, right?"

Clark shook his head. "It's not so simple. They've submitted to the *Journal of Geological Measurements*. It's not the best journal in the world, but it's respectable and has wide readership. This paper has a lot less substance than most papers in *JGM*. The reference section is very thin. From what I saw, the paper's style isn't good. The papers in *JGM* are almost all much better than what we have here. And the big thing is that most of the readership won't care about doing analysis by a hand calculator. If it weren't for its possible relevance to scientists in poor countries, I'd reject this paper in a minute. The journal has only so much space. So if this paper is accepted, a better paper will be rejected somewhere to compensate." Clark paused, shifted in his chair, and continued. "On the other hand, if the authors go to a small regional journal, published in the area of Antipodea, the paper may get lost. The libraries in undeveloped countries like Antipodea usually get only major journals like *JGM* and their own regional journals. And the databases for literature searches sometimes don't include obscure local journals. So many of the very people who need to see an article like this might not get access to it—because they might not learn about it or get it in their library."

"Why don't you just let the editor decide?" asked Leah.

"Well, she will decide, Leah. But she's a human too. If a reviewer takes a strong position, that can sometimes tip the balance on an issue like this."

◆ What should Professor Clark recommend?

CASE 12.3 Photocopying in Violation of Copyright

"Yes, Celia?" said Terence Nonliquet as he erased the board after finishing his quiz section for Comp Sci 110.

"Hi," responded Celia with slightly exaggerated enthusiasm. "I just wanted to thank you for not taking off any points on my term paper. You know, for plagiarism."

"I don't want to talk about that anymore, Celia," Terence retorted gravely.

Celia glanced around to ensure that everyone else had left the room. Then she smiled seductively at Terence. "Do you want to talk about our date?"

Terence glared at her. "There's nothing to talk about. I don't understand why you don't get it. I don't want to go out with you."

Celia hopped up to sit on the instructor's desk that separated them. "You keep saying that, Terence, but you're not very convincing. You try to hide what you're thinking, but I'm not fooled. I can tell by the way you look at me during class." She paused and adjusted her short skirt. "Especially when I wear something like this." Terence could not hold her gaze. He returned to erasing in silence.

Having gained the upper hand, Celia continued, "So I hear through the grapevine that things aren't going well between you and your girlfriend. Somebody in class said they heard the two of you arguing in your TA office." Terence tensed visibly. "Was it about me?" she teased.

"It's none of your business," Terence muttered angrily.

Celia shivered at the power she knew she was holding over him. "Really, Terence, I'm not so bad. Remember my offer. Just one date with me, and I drop your class." He continued to erase silently, eyes glued to the board. Her voice turned malicious. "Ignoring me? Well, then let's change the subject. I noticed last week you handed out a chapter photocopied from one of the texts on reserve for this class in the library. That was very kind."

Terence finished and turned around. "Oh . . . I'm glad you liked it," he deadpanned.

"You didn't happen to get the publisher's permission, did you?" Celia asked, voice dripping with poison.

"What do you care?" Terence asked guardedly.

"Well, Terence," Celia began in an exaggerated singsong. "We both know you're a man of high principle. I've always respected that! That's why you were so concerned about my copying homework from other students, and plagiarizing books, right? It just stands to reason that you got the permission you were supposed to have. I mean, publishers have a right to their royalties, right?"

"Celia, I was trying to help the class. There's a fair-use clause in the law that permits photocopying copyrighted materials."

"Yes, I've read about that," purred Celia. "People can make a few photocopies for scholarly or educational purposes. But there's no way making thirty copies of an entire chapter and distributing it free to a college class is fair use under the law." She paused and giggled mischievously. "You didn't happen to pay the royalties out of your own pocket, did you?"

"Celia, I don't see any point in talking about this."

Celia suddenly grew stern. "Oh, I do. You broke the law. And I consider it my duty to secure the legitimate rights of the publisher. I'll be writing them about what you did. . . ." She paused for effect as his face turned ashen. "Unless, of course, you can persuade me not to on that date I asked for."

"I can't believe this . . ." Terence sputtered weakly.

◆ What should Terence do?

CASE 12.4 Research Having Large Benefit for a Small but Needy Group

"There, I hope you like it," said Myra Weltschmerz hopefully as she set the freshly baked lasagna on the table in front of Martin Diesirae.

He licked his lips and rubbed his stomach. "You can count on it!" he exclaimed with exaggerated eagerness. "Your cooking is great! I really missed it since we—you know—fought. It's been about two months now. I was so happy when you offered to come over to my place and do it again."

Myra sat down with a slightly embarrassed smile. "I'm glad too." Then she paused and glanced around tensely. "I assume you've been eating out a lot," she ventured.

"Not that much. Here and there," grunted Martin between mouthfuls.

"You mean 'here and there' with Emily. . . ."

Martin paused and put his fork down. "Oh, so you heard about that. I didn't realize you knew. I mean, we haven't talked much since the fight. But you should know Emily and I aren't seeing each other anymore. Romantically, I mean."

Myra bolted up in surprise. "You're not?" she asked hopefully.

"No, we decided to cash it in. We're still working together at Tripos, and we're good friends. . . ." Martin looked intently into Myra's eyes. "But as for dating, I can see now where I've had it better all along."

Myra beamed with a radiant smile. "Martin, I'm so glad!"

Martin started to fidget at the intense emotion. "Yeah, me too." He paused, and his voice grew more businesslike. "But let's talk about the serious stuff after supper. I can't do relationship on an empty stomach."

Still ecstatic, Myra felt content to wait. "So what happened at work?" she asked dreamily.

"Busy, like always. There was something new today, though. I was talking to the chief lab chemist, Conrad. He said that a while back one of the lab team accidentally stumbled on a formulation for a metal-polishing dip that is slow, gentle, and very selective for oxide corrosion. It leaves the underlying metal intact. I had seen a magazine article last week talking about how museums clean up and preserve old artifacts. The article said that current technology still has some problems. So I showed it to Conrad, and now we think the stuff we found could be an improvement."

"Then you could sell it as a new product?" asked Myra.

"Yeah. There's a problem, though. It works only with certain alloys, so the market isn't very big. I mean, how many museums are there that do this kind of thing? The process would be complicated. It would take time and expense to scale it up to actual production, even for a small market. And our production lines were crippled by the fire a while back. Plus, we were running out of spare floor space anyway. A new production apparatus, even a small one, would need an addition to our building. That means all kinds of permits, contractors—you know, a lot of fuss and money. Tripos probably wouldn't make much profit off this stuff."

"So you have something that the museums might really be able to use, but you can't make much money from," remarked Myra. "Why don't you just sell the technology to someone else?"

"Actually, the process uses some steps that we protect with trade secrets. Even if a company agrees in a licensing contract not to use those steps for anything else, the incentive would be strong. It would be hard to police whether they adhered to the agreement."

"What are you guys going to do?"

"Good question, Myra."

◆ What should Tripos do?

SUMMARY

This unit has attempted to point out where issues of justice often arise in science and engineering. The list has not been exhaustive, and in many cases the issues have been too complex to explore in much detail. Moreover, truth and fairness as aspects of justice do not always submit to clean separation. The distinction between personal and social justice remains somewhat blurry.

Nevertheless, on a personal level truth can operate on levels of both statements and actions. Sometimes statements serve important roles in clearing up actions that are difficult to interpret. Although truth frowns upon outright deception, circumstances dictate whether spreading or withholding truth is the right thing to do. Truth by disclosure goes a long way toward avoiding problems with fairness in conflict of interest. Fairness plays a central role in assigning credit or blame for team projects (especially authorship) and in supervising employees. For supervising, fairness suggests taking a role of sponsorship toward employees.

On a social level, science and technology approach knowledge about the natural world in different ways. Whereas science demands communal ownership of truth together with disinterestedness, technology often demands the opposite. While the chief reward for creativity in science is public recognition for first publication, in technology it is the opportunity to make a profit. Patents, copyrights, and trade secrets govern the use of knowledge in technology. Public policy is crafted to balance several needs, including the free sharing of knowledge, the rights of an inventor to profits, and the provision of incentives for new ideas. Finally, we examined how knowledge should be fairly pursued, used, and accessed, and we looked at two applications: environment and paternalism.

Some Words of Caution

We have laid out only the barest outlines of some very complex ethical issues that affect scientists and engineers. While we have focused on truth and fairness in examining these questions, solving them on either a personal or social level also requires large doses of the other virtues: prudence, temperance, and fortitude. The temptation to shrug the shoulders, withdraw, or fall into simplistic thinking can loom large. As we have said, in many ways the ethical buck stops with the scientist or engineer. Usually there is no one else to take the responsibility from us. Fortunately, while the needs are great, the rewards are large.

UNIT FOUR

Advanced Topics

"I cannot and will not cut my conscience
to fit this year's fashions."
LILLIAN HELLMAN (1906–1984),
LETTER TO HOUSE UN-AMERICAN ACTIVITIES COMMITTEE, 1952

"Life is not an exact science, it is an art."
SAMUEL BUTLER (1612–1680), NOTE-BOOKS XXII

13

RESOURCE ALLOCATION

*"The bread that you store up belongs to the hungry; the cloak that
lies in your chest belongs to the naked; the gold that you have
hidden in the ground belongs to the poor."*
BASIL THE GREAT (330–379), "HOMILIES"

In this final unit, we examine some very complex subjects for which no
easy analysis is possible. These subjects come up with regularity in the
practice of science and engineering, but test the limits of ethical thought.

This chapter focuses on the allocation of resources. This allocation
takes place all the time in human living, and is governed by a principle
that moralists sometimes call "distributive justice," although "distributive
fairness" might be a slightly more precise term. Moralists have struggled
with the ethical problems of resource allocation for a long time. No general solution has emerged, and we will not attempt to construct one here.
Instead, we will simply point out some of the main lines of thinking on
this question, and indicate that circumstances and intentions make a great
deal of difference in deciding which approach to use.

What Is Resource Allocation?

Before outlining how to approach resource allocation, we need to define
more precisely what we mean. In a broad sense, many things can be distributed in some way, including credit, blame, responsibility, money, time,
and the like. However, the word "resource" usually refers to something
measurable and quantifiable that is available to several or many people.
This perspective excludes qualitative things like credit and blame. Let's list
several examples of resources, focusing on those that arise most commonly
in science and engineering:

Money: Money probably comes to mind first when most people think about resources. Businesses of all kinds need to allocate salary among employees and revenue streams among operating divisions. Funds may need to go to research and development, process upgrades, taxes, investments, and the like. Since money that is invested has a future value (because of interest or dividends) as well as a present one, allocation of funds today can have a significant impact on how they are allocated in the future. For example, a dollar spent on building renovation has a different effect on future earnings than a dollar spent on securing patent protection for a new process.

Consumables: These include water, energy, food, medicine, and virtually any other tangible item needed for human activity. In industrial practice, typical examples of consumables include process chemicals, electricity, steam, and fuel.

Time: Individuals in all walks of life need to allocate their time among career, family, and self. On the job, scientists and engineers decide how much time to give to laboratory experiments, group meetings, training opportunities, safety and cleanup, and socializing with others. This distribution involves more that just minutes or hours. It also involves the ability to offer attention and effort— most people work more effectively at midmorning on Tuesday when they are fresh than late in the afternoon on Friday when they are tired. In a more impersonal context, computer time is also in this category.

Space: Most anyone who has ever earned a living in science or engineering knows that space allocation for laboratories and offices presents a chronic difficulty in most work places. Disk and memory space on computers also falls in this category.

Services: In a laboratory context, service resources include chemical or biological analyses, secretarial and other administrative support, and technical support from machinists, glassblowers, computer consultants, and the like. In a social context, services like education and health care are resources.

"Negative resources": We have spoken so far of items to be distributed in a positive sense. We may also consider distribution in a negative sense: allocation of debt, for example. When there exists collective responsibility for a debt or other financial liability— among operating divisions of a company, for example—that responsibility needs to be allocated quantitatively. Any liability can be considered from this perspective: requirements to provide space, administrative services, and so on. Recently certain technological dangers like pollution have also come to be viewed as negative resources. This idea has been applied to air pollution, with the U.S. government selling permits to dump fixed quantities of a pollutant like SO_2 to utility companies. The limited number of such permits makes them a resource.

Given all these kinds of resources, how can we allocate them? Numerous ways have found their way into practice. As a matter of fact, sometimes a combination of methods is used. Let's look at the main methods in turn.[1]

Allocation by Merit

Allocation by merit tends to view resources as rewards. This view suggests that rewards should be distributed according to effort, demonstrated ability, or creativity. Examples in the work place are common, including job offers, salary increases, promotions, and (in difficult times) preservation from layoff. In college classes that grade on a curve with fixed proportions of A's, B's, and so on, grades become resources distributed according to demonstrated proficiency. More subtle rewards sometimes go to those who act with uncommonly high ethical standards, as with salary increases or perks for "good citizens" in the office.

However, allocation by merit breaks down for resources that are essential to human life, like food, water, and shelter. Use of merit criteria under inappropriate circumstances can lead to the trampling of very basic human rights, and tends to discriminate against those who find themselves underprivileged through no fault of their own. In impoverished countries, for example, few would argue for denying children food because they are not as productive as adults. Moreover, practical circumstances sometimes dictate that resources flow toward where they are needed most. In the work place, extraordinary resources may flow toward projects or divisions having high business priority, even when the staff perform at levels below their peers elsewhere in the company. Rearranging the staff itself may take too long or cause too many other problems.

Allocation by Social Worth

Allocation by social worth tends to take a practical view toward resources, directing them toward those who appear most likely to contribute to the common good. This view suggests that resources should move in directions that ultimately do the greatest good for the largest number of people. Criteria for social worth can include age, seniority, rank, and expertise in a crucial area. In extreme cases of war, natural disaster, or emergency, for example, social worth suggests giving more food, water, and medical attention to political and military leaders and to physicians so that these can help preserve social order and health for others. In everyday life, social worth operates at businesses when layoffs become inevitable. These cutbacks tend to preserve management and employees with the greatest experience.

As a practical matter, social worth often correlates with merit, so the distinction between these two methods sometimes remains blurred. More generally, however, allocation by social worth requires the resolution of two significant problems: determining what constitutes worth, and establishing a hierarchy among worthy things. These questions ultimately revolve around values. Establishing a hierarchy of values represents no easy task. Since people disagree over the ranking of values, they also disagree over social worth.

Allocation by social worth breaks down when the criteria for worth ignore basic human rights. For example, wealth is sometimes used to measure social worth, especially in countries with market economies. We do not refer here to mere ability to pay (which can cause similar problems, as discussed below). We refer instead to the "Leona Helmsley attitude" embodied in the infamous 1980s quote, "only the little people pay taxes." This attitude can cause food, energy, education, medical attention, and social influence to "flow uphill," thereby making severe imbalances in essential resources even worse.

Allocation by Need

Allocation by need tends to view resources in terms of basic human rights. This view suggests that every person has the same right to some minimal level of a given resource. Obvious examples include the fundamental needs for food, clothing, and shelter. This approach often operates in the wake of natural disasters, where the weakest and sickest receive the most attention. In the work world, need underlies the practice of giving poorly performing professional sports teams first choice in drafting new players. (Of course, this practice also has the healthy consequence of preventing team "dynasties" from becoming too firmly entrenched, thereby limiting overall interest in the sport.) Apart from the notion of rights, allocation by need can also have purely practical origins that make little appeal to fairness. In a business, the weak link may be a division coping with aging equipment or problems with morale. If the division forms a central component of the business's overall operations, management may need to direct considerable new resources of money and personnel to shore up the operation.

Depending on how "minimal level of a given resource" is defined, need can give license to the taking of resources from "haves" and the giving of them to "have nots." Progressive taxation systems offer good examples; tax rates increase according to income or wealth. Here, the dictates of need often run directly counter to those of social worth (when social worth is measured by wealth). Interestingly, however, both approaches have to deal with similar questions. Allocation by need requires the resolution of two significant problems: determining what constitutes need, and establishing a hierarchy among needs. As with social worth, these questions ultimately revolve around values. Since people disagree over the ranking of values, they also disagree over needs.

Allocation by need breaks down when this criterion is applied so strictly that it removes the incentive to produce. It's usually true that people work hardest when they believe they will enjoy the fruits of their labors. Those who produce the most sometimes have the fewest needs, and in such cases allocation by need can greatly reduce this enjoyment. When the most productive members of an organization put on the brakes, the entire structure suffers harm.

Allocation by Ability to Pay

Allocation by ability to pay tends to view resources entirely in terms of market forces, and in some ways represents a cross between merit and need. Merit enters in because those who produce more can afford to pay more. Need enters in because those who need a resource more will be willing to pay a higher price. Obvious examples of distribution by ability to pay include any unregulated economy, such as a black market. Earlier we pointed out how the opportunity to dump certain atmospheric pollutants has effectively become a resource in the United States, allocated by ability to pay. Some college placement offices use this principle (sometimes in combination with a lottery) by assigning "credits" to prospective interviewees, and then auctioning off rights to interview with a given employer to those who bid the most credits.

Allocation by ability to pay generally works well for nonessential items, but fails when there is not a fair distribution of wealth to start with. An ability-to-pay scheme then merely propagates these preexisting injustices. The problem becomes particularly severe when ability to pay is used to govern the distribution of necessities like food and medicine.

Allocation by Equal or Random Assignment

Allocation by equal or random assignment takes the view that no rational, unbiased way can be found to distribute resources. Equal assignment can be used for items that can be divided into very small quantities, including money, water, and food. The resource is divided into as many equal portions as there are people and distributed accordingly. Random assignment must be used for items that cannot be divided, including homes, jobs, and doses of medicine. This method generally employs some kind of lottery system that offers everyone an equal chance.

Both equal and random assignment remain free from any personal biases. However, these methods give up any attempt to account for legitimate differences in other factors we have discussed: merit, need, and so on. Equal and random assignment attempt to avoid the difficult ethical choices presented by these factors. Whereas sometimes good practical reasons exist to justify this procedure, other times the practice is grounded in moral faintheartedness.

Equal assignment breaks down when each portion of a resource is simply too small to do any good. For example, dividing a store of antibiotics into small doses during an epidemic could make each dose so small that no one benefits. Allocation by random assignment breaks down when by chance certain individuals wind up with a huge share of the resources while others receive very little. Recall that probability specifies only what will happen on average if an event occurs many times. Thus, on average all recipients in random assignment should get an equal share. Nevertheless, sometimes resources are distributed only once or a few times—for example, chances to interview with employers through college placement offices. A lottery scheme might leave some students with two or three times as many interviews as others. As another example, if jugs of water are allocated randomly in a desert on a weekly basis, and by chance some people get no water for two weeks, it matters little to their parched skeletons whether things even out on average.

Allocation by Similarity

Allocation by similarity really represents an approach to using other methods of distribution rather than being an independent method in its own right. In fact, allocation by similarity says nothing about whether deciding should be done by merit, social worth, need, or the like. Similarity merely prescribes that cases that look the same should be treated in the same way. For example, if during a desert famine we decide to allocate food on a need basis, similarity dictates that we distribute water the same way. As reasonable as the criterion of similarity may sound, its weakness lies in deciding which cases are similar and which are not. Practical situations are rarely identical, and line-drawing can become a major issue.

How to Decide among Methods

No simple rules for allocation can guarantee fairness under all circumstances. The ultimate decision depends heavily on exactly what needs to be distributed and on the specific details of each situation. Circumstances and intentions need to be considered.

Nevertheless, at least one important principle must be kept in mind when choosing an allocation method. The fundamental principle of Chapter 7 says that the obligation to avoid what is bad outweighs the obligation to do what is good. To see how this principle might operate in resource allocation, consider the contrasts between the following cases.

First consider a case in which three students decide to form a temporary lawn-mowing business to earn some money over the summer. They pool some funds to buy a lawn mower, round up some customers, and mow the lawns by turns as their schedules permit, keeping the total time spent mowing equally balanced overall. Each student agrees to an equal share of the profits, and all contribute equally to the normal upkeep of the equipment. Suppose that one day one of them is mowing some deep, thick grass and happens upon two fifty-dollar bills wedged under a weed. Should the money be shared as part of the profits? The criterion of merit suggests that the money should be kept by the finder. Now suppose the cash is in easy view, so that anyone mowing the lawn can see it. Since the one mowing the lawn is there by luck, it is tempting to invoke the method of equal allocation. Suppose further that this student is financially better-off than the other two. Allocation by need might suggest giving a bigger share to the partners. In the end, however, most people would probably agree that since the extra money is a luxury rather than something essential, the finder would not be unfair to keep the money in any of these situations.

Now consider another case involving the same students and lawn-mowing business. Suppose one day one of them is in a hurry and by accident hits a tree root that wrecks the lawn-mower blade and damages the engine. The repair costs amount to one hundred dollars. Who should pay? If none of the students has severe financial trouble, we might use the reverse of allocation by merit: allocation by demerit. The student who caused the damage was working in an unproductive way and should therefore shoulder the entire burden. However, suppose the tree root was hidden under thick grass so that even a careful worker could legitimately have the accident. If any of the students could have hit the root, equal allocation suggests distributing the cost equally. However, here the suggestion to share seems stronger than it does for the case of finding one hundred dollars, as we discussed above. The cost is a harm rather than a benefit, and the obligation to reduce harm outweighs the obligation to increase benefit. Suppose now that student who hit the root is on the verge of personal bankruptcy and can barely afford to pay tuition costs for school. The extra cost would push him over the edge. Allocation by need now suggests that the other students should shoulder the burden completely, or at least extend a loan to their poor partner. Here, the obligation to offer assistance is considerably stronger than it was earlier for the "wealthy" student to offer extra money to the other partners. Once again, reducing harm outweighs increasing benefit.

A REAL-LIFE CASE: Ethical Issues in Affirmative Action

In almost every economy, jobs are a limited resource. In the United States, the distribution policy for most kinds of work has rested on a merit basis: that is, a given job should go to the most qualified person available. The sad truth is that throughout much of U.S. history, this "ideal" policy has not matched the actual one. For many years women and various minority groups remained excluded from the best jobs, or in some cases, from any job at all. With the social advances following the civil rights movements of the 1950s and 1960s came the recognition that this situation had to change. The need for a rapid fix together with leftover barriers to employment of some groups gave rise to "affirmative action" as a policy basis for hiring. That is, applicants from underrepresented groups were to enjoy a certain measure of preference. Affirmative action became a dominant theme in the hiring practices of governmental and corporate America by the mid-1970s.

Together with the employment gains achieved by underrepresented groups over the past two decades has come a significant and divisive backlash, centering mainly around how big the "certain measure of preference" should be. As with any social debate, the issues are complex and subtle. We can only sketch some key ideas here. Opponents claim that present policies have gone too far. The most vigorous objectors argue that affirmative action as now practiced offends against fairness as badly as did the discrimination of past decades. They claim that affirmative action represents a thinly disguised quota scheme that merely perpetuates racism, sexism, and other ills. Other less vocal objectors simply state that equal opportunity has now largely been achieved, so that hiring preferences are no longer needed. These various arguments carried the day with the 1996 passage of California's Proposition 209, ending state-sanctioned affirmative action.

Proponents of affirmative action have countered vigorously. Some imply a continuing need to "make up" for the evils of the past in almost penitential fashion. Most proponents, however, focus on both anecdotal and demographic evidence that they believe proves old discrimination practices continuing to this day. They argue that affirmative action needs to remain in place until such discrimination is completely wiped out. Interestingly, people on both sides of the debate appeal to the virtue of justice (or fairness) to support their case. Unfortunately, the issue seems unlikely to resolve itself any time soon.

- ◆ Do you see evidence of discrimination today, either for or against underrepresented groups?
- ◆ If so, what sorts of policies would you put in place to solve these problems?

References

Lewis, Brian C. *An Ethical and Practical Defense of Affirmative Action.* Princeton, N.J.: Princeton University Press, 1996.

Sowell, Thomas. "How 'Affirmative Action' Hurts Blacks." *Forbes,* 6 October 1997, 64.

"With wisdom grows doubt."

JOHANN WOLFGANG VON GOETHE (1749–1832), *MAXIMS*

Note

1. This treatment follows one found in an article by Gene Outka, "Social Justice and Equal Access to Health Care," *Journal of Religious Ethics* 2 (Spring 1974):11–32.

Problems

1. Write a page or two describing an ethical dilemma involving resource allocation you have encountered in a job you've had. (If you've been lucky enough never to have been confronted with a problem like this, describe one that a friend or relative of yours has had.) Recommend what action you think you (or your friend/relative) should have taken, and give reasons for and against that recommendation. Note: you don't have to say what was actually done in real life (unless you want to)!

2. Each case below has a question after it.

   a. List the options/suboptions available to the main character who has to make a decision, together with the event tree flowing from each option.

   b. Recommend what you think the character should do.

CASE 13.1 Distributing a Limited Salary Raise Pool

"Did you hear the news, Martin?" exclaimed Emily Laborvincet as she rushed breathlessly into the office at Tripos Metal Polish. "There won't be any layoffs after all! Monica decided to sell some of her interest in Tripos to a partner, who's going to give the company some cash!"

"I know only the barest details," responded Martin Diesirae, standing up from his chair. "Who controls the company now?"

"Well, I just talked to Monica in the hospital, and she says she still has 51-percent control."

"How did she pull this off?" Martin queried. "You've been saying her mind isn't too sharp yet."

"She said she's been negotiating with this guy for a while—long before her accident," Emily replied. "The details were actually pretty well in place. She just had to decide whether to go forward."

"So does anything major change in what we're doing?"

"Not right away. We're just not broke like we used to be. In fact, Monica said there's enough money for small raises."

"Awesome!" exclaimed Martin. "Who's going to decide? Is Monica well enough? She always used to do that in the past."

"No, it's an accounting job, which means I have to do it," Emily rejoined. "Like a lot of the other stuff I do, I'll propose a preliminary plan that Monica will need to approve. I'll have to justify everything I suggest. But so far she's never changed anything I've done."

Martin grew uneasy. "How much money are you talking about?" he asked.

"The raise pool is big enough for an average of about 3 percent for our fifty-one people. Monica suggested that everyone get at least some raise because of everything we've all had to suffer through recently—absent management, fire in the factory, and the potential for layoffs. But she said that wasn't cast in stone."

Martin drew a deep breath. "Great. So you're recommending my raise, too. Now that we were dating and you dumped me . . ."

"Martin, I don't like the word 'dump.' I don't think it was like that. And I said I wanted to be friends. Anyway, I hear you and Myra are back together again. I think it was all for the best. As far as a raise, everyone knows you've worked as hard as anyone here these past few months. You've got as much claim to a good raise as anyone."

"Maybe you should just give everyone the 3-percent increase. That makes things simple," Martin observed, somewhat relieved.

"That would be simplest," Emily agreed. "But I hear Conrad has been unhappy ever since the fire. He was especially mad at the possibility he might be laid off, that we needed a quality control chemist more than his development skills. Conrad heads our development lab, and losing his creativity and experience would really hurt us. Probably he of all people should get more than average."

"Yeah, and I hear a couple of the floorworkers have been calling in sick a lot lately," commented Martin. "It's pretty obviously fake. And they've been sloughing off since the fire because they think no one's watching. They shouldn't be getting much of a raise."

Emily nodded. "The problem is, we don't have a formal review procedure. Monica just made judgments from her gut, which I guess is fine when you own the company and you have a long history with the place. But I don't own it and I've been here less than a year. I don't know everyone that well, especially on the floor, so I have to get some input from the floor supervisors. And Monica says she wants this done in two weeks because that's normally when raises are handed out. She doesn't want any delay. There's not time to put in a real formal review procedure."

"Good luck," chortled Martin.

◆ What should Emily recommend?

Case 13.2 Individual Credit in Team Projects

"I never expected organic chemistry to be like this!" exclaimed Emily Laborvincet to her boyfriend Todd Cuibono as they left their lab class at Penseroso University.

"Yeah, Professor Isolde is a piece of work," retorted Todd. "Who ever heard of having students distribute points among themselves for a lab report?"

"And she didn't tell us till all the groups were done with the project! Three weeks we've worked on this synthesis. That's bad enough. All those steps, and all that characterization. Ugh! But then to have her tell us this!" Emily shook her head in disgust.

"So how does the scheme work again? It was sort of complicated," observed Todd. "We hand in a joint report next week. The reports are graded on a numerical scale from 0 to 300, right?"

"Uh, huh," Emily broke in. "Three quarters of those points get distributed evenly among the three people in each group. But for the remaining quarter, we're supposed to decide among ourselves how to distribute them. We report that distribution to Isolde on a paper signed by all the group members. And if we can't agree, she penalizes the group by taking half the points and assigning them equally. The other half get thrown away."

Todd rolled his eyes. "What a nut case. I'm all for giving people who work hardest within a group the most credit, but I don't think this is a good way to do it." He stopped walking and looked gravely at Emily. "We're going to have trouble with Salina, you know."

Emily nodded silently. After a moment, she responded, "I know. She hasn't done much of anything. I think she spent half the period today talking to people in other groups. And she doesn't understand the science. I thought she would make a good partner, but I was fooled big-time."

"If we could get her to write up the report, at least it would make the work load more even," Todd ventured. "All the calculations, graphs, and text will be a big job to put together. But you're right. I don't think I'd trust her to do it right even if she agreed to the task. But maybe that's OK. Maybe she'll agree to claim only a few of the extra points."

Emily shook her head. "I doubt it. You remember when I went to the bathroom toward the end of the afternoon? She happened to go at the same time. In a roundabout way I tried to find out what she thought about the whole grading thing. She said she thought it was great, and we should just divide the points evenly because that's what a lot of other groups were doing. She did apologize for not doing more in class, though, and offered to write most of the final report."

Todd scratched his head. "Great. Then there won't be many points to divide. It's hard to see a solution that doesn't give her a free ride. If we write the final report, she should get almost no extra points. Then from her point of view, she'd be better off not to agree to anything we propose, and just take a sixth of those points after Isolde's penalty and the equal distribution."

"Let's worry about it tomorrow," suggested Emily. "I'm too tired to think about it now."

◆ What should Todd and Emily do?

Case 13.3 Distributing Limited Computer Time among Students

"You know, Leah? Professor Bligh really gives me headaches," murmured Terence Nonliquet to his girlfriend Leah Nonlibet as he looked up from the computer screen in front of him. He paused for a moment, then continued. "He assigned this computational homework problem that's way too difficult for a beginning course like Comp Sci 110. So I've had students flooding my office hours. And the problem requires them to use a special software package on the Engineering College's mainframe computer instead of the campus's personal computers. That means the students have to use timesharing, and get limited run time. Well, because the problem is so hard and the students don't know what they're doing, about two thirds of them ate up their time allotment before finishing. I complained to Bligh about this along with the other TA's, so Bligh got some more computer time and told us to distribute it as we see fit. But that's not an easy task. I've got to figure out an algorithm for my section that's both fair and manageable."

"I don't see the big problem," Leah retorted. "Just distribute it equally to all the students."

"Yeah, but some of them already finished. They don't need more time. The homework problem was supposed to give experience with this package, not test how quickly people could develop working code. I want to give the time to the people who need it." Terence pointed to a class list in front of him. "I asked the students with trouble to see me after class. This list shows who's not done and how much time they think they'll need. The trouble is, the sum of what they need is about 30 percent more than what I've been given to distribute."

Leah reached over and took the list. She examined it carefully, then looked up at Terence, who was lost in thought. "Two people asked for about twice as much time as the others," she observed. "What are you going to do about them?"

"Oh, they're way behind for some reason. Anyway, I'm thinking of just cutting everyone's request by 30 percent to match the time I have. That way everyone will get most of what they need."

"But that just rewards people who are stupid or inefficient," retorted Leah with an edge in her voice. "Why don't you just distribute the time equally among all those who ask?"

Surprised at her tone, Terence frowned and sat upright. "You don't have to get mad," he said. "I was looking for advice, not more headaches."

Leah pointed at the paper and remarked in measured tones, "Well, it seems one of the students most hungry for time is a certain 'Celia Peccavi.' I seem to remember that name." Leah's voice grew more accusing. "I asked you about her a couple of months ago. We were in this office, and you had just gotten off the phone with her. You told me you never had her in class before this semester, but I happened to see your grade roster from last semester a week later when you left it on your desk. Her name was there." Leah paused for effect, and looked Terence dead in the eye. "You lied to me. Why?"

Terence's glance fell to the floor, and his face reddened with embarrassment. He shrugged weakly in silence. After a moment, he looked up gloomily and tried to head off the impending storm. "It's not such a big deal. I mean, what difference does it make who was and wasn't in my class? And what does that have to do with distributing computer time?"

Leah stood up and glared down at Terence. She refused to be put off. "It was a big enough deal for you to lie to me! I think it's because something is going on between the two of you behind my back. You told me no one in your class ever tried to make a pass at you. But I heard through the grapevine that she invited you to a volleyball game last semester and you went. That doesn't sound to me like a typical TA-student relationship!" Terence stared ahead in silence. Leah paused, then continued relentlessly. "So I think it's more than coincidence that the algorithm you've devised for handing out computer time gives her a big benefit."

This last shot infused new energy into Terence. "That has nothing to do with it!" he contended. "My method makes sense. I'm distributing on the basis of need. I think you're just paranoid. And who would guess that you'd want everything so equal? You weren't so interested in equality when we went to that Chinese place! Remember? With the big group? The food was too spicy for you, so you didn't eat much. I calculated the bill, and was going to split it equally among everyone. But you didn't want to pay much of anything because of what you ate. You wanted me to charge everyone else for your share! I wouldn't do that, and wound up swallowing both your bill and mine, the way I always seem to do. I think you're being hypocritical."

"I'm not hypocritical," Leah shot back. "And I certainly don't lie when my 'special somebody' asks me a simple question!" She grabbed her jacket off a nearby coat rack and stormed out.

♦ How should Terence distribute the computer time to his students?

Case 13.4 Distributing Grades among Sections

"Yeah, Veronica, I'm thinking the same thing. We have to get this decided before the semester ends. It'd be more fair if we normalized just the homework scores," said Terence Nonliquet into the phone as he sat in his TA office. "I mean, the exams are graded by all the TA's collectively, while the homework for each section is graded by the TA for that section. . . . Uh, huh . . . yeah, I know. Some of the TA's grade easier than others. But not me. If anything, I'm pretty strict. And by luck I happen to have a set of really good students. I don't have the results of the final homework yet. But I can tell you how things are shaping up now. While my students' homework average is about the same as the class average, they're actually doing better. . . . Uh, huh. . . . Yeah, and that's not all. I looked at the exam averages for my section, and it's 3 points above all the other sections. So my students get ripped off with Professor Bligh's policy. Yeah, I tried to tell that to Bligh, but as usual he wouldn't listen. He kept saying that the classes are big enough, so there should be no significant differences in student quality—that differences in exam averages come from how good the TA's are, which the students can't help. . . . Yeah, I'm thinking of ignoring him and normalizing just my homework grades to the course average when it comes time to calculate the semester grades. You too? Good! If we can just get the other TA's to agree, that'll be great. Bligh doesn't need to know . . ."

Terence looked up and saw his student Celia Peccavi standing at the doorway to the office. His voice became hushed and hurried. "Look, Veronica, I have to go. Can you call the other TA's? They'll probably listen to you more than me because you're a grad student and I'm only an undergrad. You will? Thanks! Gotta go." With that, he hung up. Looking up at Celia, he exclaimed, "What are you doing here? It's 9 at night! My office hours were yesterday."

Celia grinned broadly. "I just wanted to know if you knew what my last test grade was yet. I saw the light on in the office from outside, and figured you'd be in."

"The grades aren't done yet," responded Terence testily. "I told you and everyone else in the class it would take at least until tomorrow, and maybe longer."

"Well, anyway, I'm glad you're trying to do the right thing—you know, with the grades. Right now I'm on the border between a C and a D for the semester. It didn't help that I bombed that big homework assignment . . . you know, the one where you gave out extra computer time because so many people used theirs up? You asked the people who hadn't finished how much extra time they would need, but then wound up distributing it equally among them. What I got wasn't even close to what I asked for, so I didn't finish."

"I can't help that," Terence retorted dryly.

"I know," Celia continued. "But now what you're doing will help me a lot." She paused and smiled again. "And I know you always have the right thing at heart. That's why I didn't call the publisher of that book—you know, the one you copied a whole chapter out of and gave to the class?" She shook her finger at him playfully. "You violated copyright law, you know, by not getting permission and paying royalties! But I think you meant well."

"Celia, I'm too busy for this . . ."

At that moment, Terence's girlfriend Leah Nonlibet appeared at the doorway next to Celia. Terence turned pale. Noticing this, Celia eyed Leah shrewdly. Leah glared first at Celia, then at Terence. "So what's this?" she barked. "I stop by to drop off a late night snack—just to be nice—and find this 'Celia' in your office. Don't you think it's an odd time to have students chit-chatting in your office—at 9 o'clock?"

Terence was too surprised to reply, but Celia had no loss for words. "Hi, I'm Celia," she said with pretended cheeriness, extending her hand toward Leah for a handshake. "I don't think we've met. Do we know each other?"

Leah glanced at Celia's hand with disdain and refused it. "Terence, why is this girl here?" she demanded.

Again, Celia broke in before Terence could respond. "I'm in Terence's class. I was just asking about my grade." She sashayed over to where Terence was sitting, and stood next to him. "Terence was telling me how he could cook things so everyone in his class could get a better grade—especially me, since I'm a borderline case." Her voice dripped poison. "Isn't that sweet?"

This remark jolted Terence out of his near trance. He stood up and pushed her away. "I'm not cooking grades, and I'm not doing anything for your personal benefit!" he cried. His face turned livid. "You never stop harassing me. Will you leave me alone? You're not supposed to be here." He pointed to the door. "Now get out!" Startled, Celia stood motionless, looking perplexed. "You heard me!" Terence roared. "Get out! Don't ever come here again!" Wordlessly, Celia slunk past Leah out of the office.

After a long pause, Leah held out the bag with the snack she had prepared. "Well, here are some goodies if you want them," she ventured.

"Thanks," he muttered. "I'm not hungry now, but maybe I'll eat them later. I'd really like to be alone now, if you don't mind."

"Sure," she said quietly. Leah left without another word.

◆ What should Terence do about the grade normalization?

14

RISK

❖

"A man is truly ethical only when he obeys the compulsion to help all life
which he is able to assist, and shrinks from injuring anything that lives."

ALBERT SCHWEITZER (1875–1965), *THE PHILOSOPHY OF CIVILIZATION*, PART II

The present chapter focuses on risk in the practice of science and engineering. As with any human endeavor, scientists and engineers have the potential to bring about both benefit and harm by what they do. Many risks appear on the small-scale, person-to-person level. Examples that threaten typical process operations include poorly trained personnel, sloppy housekeeping, and use of production machinery outside design limits. Other risks appear on the large-scale, social level. Examples that threaten large groups of people include acid rain, global warming, and highway congestion.

A Historical Perspective

Risks on the personal level have plagued science and engineering from their earliest days. Early alchemists often sickened from prolonged exposure to the toxic substances they made. In 1753 the Russian experimenter Georg Wilhelm Richmann met his death by electrocution in experiments with lightning. On the other hand, risks on the social level have burst into widespread public awareness only during the past few decades. The benefits of science and technology for the world community are very clear—improved health and longevity, faster communication and travel, and higher living standards, to name just a few. In fact, during the 1800s and much of the 1900s, these benefits showed themselves with such force and intensity that by the 1950s and early 1960s many people expected (at least in the United States) that technology would largely do away with disease and household labor. In universities, departments of leisure studies and related disciplines sprang up to examine what people would do with all their well-being and spare time. Unfortunately, the energy and environmental crises of succeeding decades put an end to these overly optimistic projections with a resounding bang. The exploitation of technology for military purposes and for snooping into people's private lives has become increasingly frightening. Indeed, uncontrolled use of technology has shown an ugly, dangerous, and dehumanizing side. Put another way, it has become clear that many technologies carry significant risks for harm to individuals and societies.

Moralists have struggled with the ethical problems of risk for a relatively short time. Risk assessments necessarily include concepts from probability theory, a field that came into being only during the 1600s.[1] Thus, classical moral theories that originated much earlier did not incorporate these ideas. Since risk has become a front-burner social issue only during the last three to four decades, philosophical and moral treatments of the subject remain relatively new and untested. No generally accepted approach has emerged, and we will not attempt to construct one here. We will simply point out some of the main lines of thinking on the question.

Defining Safety and Risk

Discussions of risk sometimes take place in the context of "safety." "Safety" is an abstract term with many meanings; we can speak of safety on physical, psychological, and economic levels, for example. "Risk" is similarly abstract, and in common speaking refers to almost any potential threat to safety. However defined, both safety and risk have significant ethical components. As we have pointed out several times in this book, reasoned moral discussion requires a common and fairly precise understanding of what key terms mean. Hence, we will dwell at some length on defining our terms. Since the concepts of safety and risk complement each other, for brevity we will focus mostly on risk.

As we indicated above, risks can threaten people on many levels. However, in this elementary treatment we will restrict our attention to physical well-being.[2] Our understanding of what the word "risk" means is very important, because it is well known that scientists and engineers tend to define the word differently than the public at large. Many of those who have technical training tend to understand risk very narrowly as simply the probability that some given harm will come about. For example, risk might refer to the probability that a particular level of carcinogen causes cancer in a large population. Laypeople, on the other hand, tend to incorporate moral importance into their ideas about risk. Thus, the risk that something bad will happen includes both probability (very loosely defined) and moral importance.[3] Furthermore, laypeople often fail to distinguish risk as separate from the acceptability of risk. Risks tend to be seen as unacceptable if they are difficult to understand, unfairly distributed, or not within direct control.

Both perspectives about risk remain incomplete but contain real wisdom. Those with technical training better understand quantitative probabilities, but sometimes ignore the broader aspects of value judgment. This perspective comes across as mechanical and sterile. Those without technical training better understand how moral importance and personal control enter into a decision, but sometimes confuse attitude with probability. This perspective comes across as inconsistent and irrational. Unfortunately, in some cases the chasm is too large to bridge in a timely way,[4] causing real problems for engineering projects.

Evaluating Risk

Engineers often analyze complex automated systems in terms of event trees or fault trees. An analysis using event trees starts by assuming that a particular component will fail, and traces through the consequences of that failure for the entire system. A complete treatment requires an estimate of the probability for the initial failure and for each consequence. Fault trees trace the process in reverse. The analysis begins with some undesired event at the system level (for example, an airplane crash) and traces back to the component failures that could have led to that consequence. Again, a complete treatment requires estimates of probability for each step. While these forms of analysis can prove very useful as systematic aids to assessing risk, they have severe limitations. No one can possibly imagine all the ways a complex system can fail—in particular how circumstances can conspire to produce failures that might not occur otherwise. Furthermore, estimating the probabilities required at each step sometimes proves to be nearly impossible, since the needed data cannot be obtained. Overemphasis on event trees or fault trees can lead to a mind set that prepares only for anticipated problems, not leaving enough room for unforeseeable difficulties.

Scientists often find themselves deeply involved in risk assessment as well. For example, epidemiologists and other life scientists attempt to develop models for how foreign substances act in the human body. Some toxins, carcinogens, and radioactive substances seem to be harmless when present below a certain threshold level. Other substances seem to follow a linear, no-threshold model, meaning that the probability of harm exists no matter what the level. Deciding between these models often stirs up great controversy. Contributing to the debate is the fact that low levels of foreign substances tend to cause illnesses or other problems at levels that epidemiologists consider to be "in the grass," meaning that the effects cannot be distinguished easily from the natural incidence. Nevertheless, whatever conclusions are reached have crucial implications for public policy because many analytical chemists devote their careers to developing more precise and sensitive ways to measure low levels of trace compounds. Improved sensitivity has little effect on risk models that employ thresholds, but has considerable impact when risks remain present all the way down to zero exposure.

In many public and corporate policy decisions, options are chosen largely on the basis of a cost-benefit or risk-benefit analysis. A cost-benefit analysis assesses each option by adding up the dollar value of each possible consequence (positive for benefits, negative for harms) multiplied by its probability. Some analyses even incorporate the time value of money due to interest. For example, if the monetary benefits appear immediately but the harms do not show up until later, the future dollar amounts can be compared with the present ones by assuming a suitable interest rate. A risk-benefit analysis operates like a cost-benefit one, except that the units need not be monetary. The unit of measure for each consequence could be deaths, injuries, or other countable item. Of course, we have emphasized throughout this book that many dimensions of moral analysis do not submit to quantification. Nevertheless, despite their limitations, cost-benefit and risk-benefit analyses serve a useful purpose by providing a decision process that remains open to examination and modification. Moreover, once the decision has been made to start a project, even apples-to-oranges comparisons of consequences serve important uses when comparing among similar project designs.

Laypeople also balance benefits against harms. For example, one study has shown that on average, people are willing to accept a probability of dying in the work place that increases with the cube of their wages.[5] Nevertheless, the process seems to involve many factors other than those included in standard risk-benefit analysis, including the degree of control people have in choosing, the way information is presented, the aversion to harm versus the desire for benefit, and proximity to the danger. Let's examine these factors more closely.

Many studies have shown that people are far more willing to accept risks they can control than those they cannot. For example, one study has shown that on average, people will accept a probability of death from voluntarily hunting, skiing, or smoking that is *one thousand* times what they will accept from involuntarily living in areas with frequent natural disasters or a nuclear power plant.[6] Although some of this difference may come from misinformation about the death rates involved, most of it appears to arise from the degree of control in deciding.

The way information about probability is presented also affects the perception of risk. In one study,[7] people were asked to choose between two options for fighting the outbreak of a rare disease that, if left untreated, was expected to kill six hundred people. The question stated that treatment option A would definitely save two hundred people, whereas option B would have a one-third probability of saving all six hundred but a two-thirds probability of saving none. Whereas 72 percent of the sample chose the first option, only 28 percent chose the second. Then, a second sample was offered the same question with the options worded differently.

The second question stated that treatment option A would definitely allow four hundred people to die, whereas option B would have a one-third probability that no one would die but a two-thirds probability that all six hundred would die. This time, only 22 percent chose the first option, and 78 percent chose the second—even though nothing objective had changed in the cases!

This study not only showed that the presentation of information matters, but also agreed with the observation by others[8] that people tend to be more willing to submit themselves to take chances to avoid harms than to preserve benefits. More generally, people tend to exert more effort to avoid a loss than to secure a gain.

Proximity to danger also affects willingness to accept the possibility of harm.[9] People tend to feel more threatened by a harm that affects friends and relatives than by the same harm that affects strangers, even when the chance of suffering the harm is greater for the strangers. Similarly, harms expected with great likelihood in the remote future seem less threatening than less likely harms expected soon. This effect seems to come partly from an attitude of "out of sight, out of mind" together with a vague belief that the passage of time may produce some unforeseen remedy.

Scientists and engineers need to be aware of this psychology in the public they serve. Failure to appreciate these factors by those who deal directly with the lay public will certainly lead to misunderstanding and ultimately mistrust. However, even those who remain at a distance from the public cannot ignore these behavior patterns, as otherwise products can be mistakenly designed or marketed in ways that inflame the public's concern over technological risks.

Making Decisions about Risk

Regardless of how risk is perceived, the fact remains that much of what science and technology produce contains risks to human health. The risks may be inherent in the products themselves, or may arise through the production process. Whether on the personal or social level, two burning questions arise about deciding in the face of risk.

What Criteria Should Determine the Acceptability of a Risk?

Both truth and fairness contribute to judging the acceptability of risk. Truth ideally requires that all those affected by the risk know about it, understand it, and give their consent to it. On the person-to-person level involving workers with technical training, this ideal often represents a legitimate practical goal. On the social level involving larger numbers of people, many of whom have no technical training, true informed consent may remain out of reach for practical purposes. However, this limitation does not imply informed consent can be ignored, or that efforts toward it should be abandoned or scaled back. Fairness also enters into the decision regardless of whether informed consent is achieved. We must consider the sum total of benefits and harms together with their distribution. If some people must shoulder a disproportionate share of the potential harms without a corresponding share of the benefits, this fact needs to be accounted for.

Who Should Determine the Acceptability of a Risk?

Assigning ultimate decision-making power requires that we answer two questions. First, who decides whether the available data provide an adequate basis for determining risk acceptability? For example, on whom should the burden of proof fall—producers of technology or consumers? Second, if there is an adequate basis, who makes the final decision? For example, when there is disagreement over the interpretation of data or the ranking of moral values, do we give the benefit of the doubt to those promoting a new technology or to those opposing it?

On a person-to-person level, risk can be approached in several ways. In the team environment of the work place, team leaders or supervisors generally have the responsibility for decision-making. Among equals without a designated head, some form of majority rule usually works, with the size of majority needed for action increasing with the gravity of the consequences.

On the social level in a representative democracy, the situation becomes far more complex. In a market-oriented economy, most new technologies are introduced by companies in order to make a profit. Given the short-term pressures of the competitive marketplace, producers can be tempted to introduce products regardless of the risk to public health. This conflict of interest can pit producers against the public users. Moreover, producers usually understand new technologies much better than users, creating problems for informed consent. Serious tension between technical experts and the lay public comes into play. Although the government usually has less conflict of interest than a producer does, the government often cannot maintain the same level of expertise for a particular technology that the producer has.

Some General Guidelines

This chapter has raised many questions. Some have no general answer, but require more detailed examination of the details of circumstances. However, certain guidelines for handling risk operate in almost all cases.

First, we may tolerate increasing risks only when the possible benefits of what we do increase proportionately. In other words, we cannot take large risks to secure small benefits.

Second, in work that involves others we must work diligently to obtain informed consent. As we indicated earlier, fully informed consent is sometimes very difficult to get. In such cases, we must ensure proportionality between our efforts and the degree of risk. If the risk is very large, we need to work that much harder to secure informed consent.

A REAL-LIFE CASE: Experimental Drug Testing in Humans

For a medicinal drug to be proven safe and effective, it must be tested in people. Currently such testing is done by pharmaceutical companies and doctors under the close supervision of the Food and Drug Administration (FDA). Prior to FDA approval, a proposed new drug must undergo three phases of testing. In Phase I, the drug is given to a limited number of healthy volunteers to check for serious side effects. In Phase II, the drug is given to a group of sick volunteers to check for effectiveness and additional side effects. A control group receives either a placebo or, in the case of a very serious disease, the best available alternative treatment. In Phase III, the drug is given to a much larger population of sick people to obtain a better reading on effectiveness, interactions with other drugs, rare side effects, and the like. FDA oversight is needed for this process because of the enormous conflict of interest drug development unavoidably poses for the manufacturers. Whereas drug companies no doubt want to make sick people healthy, the companies also want to make a profit. The temptation to rush a new drug to market without adequate testing is simply too great, partly because the costs of developing and testing (in the hundreds of millions of dollars) are so enormous.

However, problems remain even with the most careful oversight. In the end, the trials approach humans both as people needing treatment and as abstract subjects for research. Medicine and research do not always recommend the same actions. For example, research concerns itself with measuring the effectiveness and risk of a drug against a carefully specified control group. Such experiments take a great deal of time, and tend to lengthen the approval process. Treatment, however, concerns itself with healing as many people as possible as quickly as possible. Thus, as soon as experimental trials show significant promise, there may be good reason to begin giving the drug to the control group and to distribute it to as many people as possible. These actions shorten the approval process, but increase the risk that unforeseen negative effects will crop up. Right now, current approaches tend to tip the balance toward "treatment" in cases of terminal illnesses like AIDS or heart disease, while the "research" approach finds more rigorous use with less critical illnesses.

Another key issue involves informed consent. Some drugs have very sophisticated modes of action carrying poorly understood risks that are difficult to explain to volunteers. Adequate safeguards need to be put in place for children, prisoners, and the mentally ill. Some opponents of human drug tests believe such safeguards are impossible to realize in practice, and worry that researchers concerned with advancing their careers or securing grants will skew the direction of the research away from the best interests of the patients. FDA policies in these areas continue to evolve.

◆ Do you approve of drug testing using children or the mentally ill as subjects?

◆ What should be done about people with life-threatening illnesses where new drugs may provide the only effective treatment?

References

Flieger, Ken. "Testing Drugs in People." *FDA Consumer* 28 (1994): 16–19

King, Nancy M. P. "Research on Humans: What Have We Learned?" *Forum for Applied Research and Public Policy* 12 (1997):132–136.

"...all men who deliberate upon difficult questions ought to be free from hatred and friendship, anger and pity."

CAIUS JULIUS CAESAR (100–44 B.C.), QUOTED IN C. C. SALLUSTIUS, *THE WAR WITH CATALINE*, I, 1

Notes

1. Probability theory was first inspired by Renaissance gamblers seeking advantage at games of cards and dice. Tartaglia and Cardano both offered clever analyses of gaming problems. But their work seemed too greedy for respectable mathematicians and too technical for gamblers, and so it lay largely forgotten. Probability theory as we know it began around 1650 during a shared coach ride between the French nobleman Chevalier de Mere and the philosopher/mathematician Blaise Pascal. The two debated the ancient problem of how to split the pot in a game of dice that must be discontinued. Pascal pondered the problem for several years, and then posed it to the French jurist and parliamentarian Pierre de Fermat. The subsequent research of these two gave birth to the new field of study.

2. For an interesting treatment of the broader risk that technology poses to the way the human species views itself, see C. S. Lewis, *The Abolition of Man* (London: William Collins Sons, 1946). A more recent treatment appears in James F. Childress, "The Art of Technology Assessment," *Priorities in Biomedical Ethics* (Philadelphia: Westminster Press, 1981), 98–118.

3. Recently some technical ethicists have begun to view risk explicitly as the product of importance and probability. See, for example, William R. Lowrance, "The Nature of Risk," in *Societal Risk Assessment: How Safe Is Safe Enough?*, Richard C. Schwing and Walter A. Albers, Jr., eds. (New York: Plenum, 1980), 6.

4. See Paul Slovic, Baruch Fischhoff, and Sarah Lichtenstein, "Risky Assumptions," *Psychology Today*, June 1980, 44–48.

5. Chauncey Starr, "Social Benefits Versus Technological Risk," *Science* 165 (1969):1232–1238.

6. Ibid.

7. Amos Tversky and Daniel Kahneman, "The Framing of Decisions and the Psychology of Choice," *Science* 211 (1981): 453–458.

8. See for example William D. Rowe, "What Is an Acceptable Risk and How Can It Be Determined?," in *Energy Risk Management*, G. T. Goodman and W. D. Rowe, eds. (New York: Academic, 1979), 327–344.

9. See Mike W. Martin and Roland Schinzinger, *Ethics in Engineering*, 3rd ed. (New York: McGraw-Hill, 1996), 132 ff for a further discussion. Interestingly these authors follow reference 8 in plotting a harm-benefit function that has thresholds near the neutral point of small harms or benefits. The lack of effort below the threshold on the harm side supposedly comes from the human propensity to ignore small harms in order to avoid anxiety overload. Martin and Schinzinger postulate a (smaller) threshold on the gain side that represents a combination of inertia and "generosity" that prevents people from instantly seeking selfish gain. These thresholds are said to vary with the circumstances of individuals.

Problems

1. Write a page or two describing an ethical dilemma involving risk you have encountered in a job you've had. (If you've been lucky enough never to have been confronted with a problem like this, describe one that a friend or relative of yours has had.) Recommend what action you think you (or your friend/relative) should have taken, and give reasons for and against that recommendation. Note: you don't have to say what was actually done in real life (unless you want to)!

2. Each case below has a question after it.

   a. List the options/suboptions available to the main character who has to make a decision, together with the event tree flowing from each option.

   b. Recommend what you think the character should do.

CASE 14.1 Welfare of Others versus Recreation

"I'm so glad to hear you and Martin are back together again," said Dolores Sola enthusiastically to Myra Weltschmerz as the two cleared the dinner table in Dolores's kitchen. "For a while I didn't think it would happen, especially with him dating that other woman."

"Uh-huh." Myra agreed. "I was pretty much a complete mess for six weeks. I'm glad he stopped dating her. We're still a little fragile, though—not totally out of the woods."

"Anyway, with your getting back together with him, and the warm weather here, I bet there'll be more spring in your step. I know the longer days help me. If nothing else, I can send my kids outside."

"I thought the warm weather made that mulch pile near your window stink," observed Myra. "You were thinking of moving because of that."

"I know," replied Dolores. "I still don't like it, but it's just too much work and upset to move. I'm going to renew my lease when it comes up."

Myra finished wiping the table. She turned more somber. "Dolores, I have to ask you something. When I was baby-sitting this evening, I brought some candy for Garrett and Lorelei—filled chocolates. We have a pattern now. I always bring eighteen pieces, half caramel and half strawberry. Garrett gets ten because that's how old he is, and Lorelei gets eight for the same reason. But Lorelei gets all strawberry, because she doesn't like caramel. We've been doing this for weeks. But today Garrett threw a fit when we did the usual split. He wanted more candy, and more strawberry. There was no reasoning with him. I started to get fed up, and told him so. Then he said, 'What are you going to do? Smash my face? Go ahead! I can take it! Mom does it!' Dolores, you don't really beat him up like that, do you?"

Dolores turned ashen and sat down, burying her head in her hands. She sat that way for a long time. Then she murmured, "I don't mean to. I love him, really. Sometimes he just drives me crazy. When I'm tired, I lose control sometimes. Not very often, but sometimes. Like last weekend. He was calling Lorelei names, and wouldn't stop. I lost my cool, and slapped him in the face. Really hard. Then I started swearing at him, and couldn't stop. I can't believe I did it. Later I cried, and begged him to forgive me. That's what I always do when it happens."

Myra sank down into her own chair. "Dolores . . . that's wrong. . . ."

Dolores looked up, tears in her eyes. Her voice quavered. "I know. But it's hard raising them alone, on a salary that's nothing. I just need a break. I was going to ask you, before you left. I need you to watch them one other day a week for two or three hours. Any day you want, as long as I'm not working. Then I can do the shopping, and just clear my head without them around. And the kids don't like my other sitters."

Myra grew nervous. "Dolores, we had an agreement. You remember . . . we made it right after Martin blew up at me. I promised I would do Tuesdays and Thursdays unless I was sick. Other times if I could, but nothing else regular. Now you want me to do more on a regular basis."

"Not every week," responded Dolores. "Maybe every other or every third one. Can't you see? I need you. Desperately. If I don't get re-lief, I might really hurt one of the kids some time!"

"I know you need help, and I like your kids. But I need time, too. Remember, it was my baby-sitting for you that drove Martin crazy to start with."

"Myra, please!" Dolores wailed.

"I'll have to think about it," Myra responded vaguely. "I have to go now."

◆ What should Myra do?

CASE 14.2 Welfare of Company versus Vacations

"You wanted to talk to me?" asked Celia Peccavi as she walked into the manager's office at Pandarus Pizza.

"Yeah," replied Thorne Mauvais. "Celia, you know how you've been pushing for more hours for the past month? Well, I can finally accommodate you. Beginning with summer break in two weeks. We've got two people quitting as of then. You're working three nights a week Monday through Friday now, if I remember. I need you for all five week nights."

Celia's eyes widened. "Starting in two weeks? I can't do that. When classes end I plan to take some little out-of-town trips on the days I don't work. With some friends. I won't be here in the evenings on those days. The week after is OK, but not then."

Thorne grew irritated. "But Celia, it's in two weeks that I need you. I've got a couple of other good candidates lined up after that week. But two weeks from now is the pressure point. I was count-ing on you, since you were pushing so hard for the hours. I thought you'd be happy."

"Thorne, I am—sort of." Celia paused. "I really do want the extra work. I can use the money. I just don't want to blow off my friends after we planned everything unless there's a really good reason." She paused again. "If you pay me overtime—time and a half—on the two extra days, I might be persuaded."

"Overtime? That's ridiculous!" Thorne snapped. "What is this, ex-tortion?"

Celia stepped back slightly, shocked by this outburst. She tried her best to answer mildly. "Extortion? Thorne, I'm trying to help you out. It just has to be worth my while. I have a life outside Pandarus Pizza, you know."

To Celia's dismay, Thorne began to storm. "Look, if you can't ac-commodate me, I'm not going to accommodate you. You're on thin ice with me anyway, after yelling at me a while back in front of everyone. And Todd Cuibono told me you've been giving away food to your friends without their paying. Plus he says you give him nothing but trouble, arguing with him all the time. You should be thankful you have a job at all, let alone get extra hours!"

Celia's face reddened with anger, and she counterattacked. "Thorne, that's garbage! I yelled at you because you were letting the chef scrape sausage he dropped off the floor and put it into the pizza. You're lucky I didn't call the health department! And you know Todd hates my guts. He'll say anything to get me in trouble. And he's the one who starts those fights, not me! I don't have to let him step on me! You know, if you had been nice to me just now, I might have been willing to deal. But now I don't think so."

Thorne jumped up from his chair behind his desk. He shook his fist. "I've taken enough from you! If you can't work in two weeks all five days, I'll see to it that your hours never increase. And I'll do my best to cut them!" he snarled.

"And how will you carry on when you're short of two people?" Celia shot back viciously. "You can't operate this place like that! It's impossible! The food and service will be junk." She paused, her face contorted with rage. "Or will you just scrape food off the floor again?"

"Get out of here!" Thorne shouted. "You'd better be here on those two days! And if you try to go over my head to the owner, I'll see that you're fired!"

Celia turned wordlessly and slammed the door behind her. Although sanitary conditions had improved since she had scolded Thorne about them in the sausage incident, they were still not perfect. Celia strongly suspected they would degrade again in the short-handed kitchen if she were not present.

◆ What should Celia do?

CASE 14.3 Informed Consent

"You asked to talk to me, Conrad?" said Martin Diesirae as he knocked on the office door.

Conrad, the senior development chemist at Tripos Metal Polish, looked up and nodded. "Since you're in charge of our customer relations these days, I needed to bounce something off you. . . ."

"You mean about that new metal dip you developed—for museums?" Martin broke in. "I really don't think we ought to pursue that. The company has just been through too many changes."

Conrad smiled at Martin's earnestness. "I agree with you, although that's not what I wanted to talk about right now. I'm thinking about something different. It's about a reformulation of our APZ polish for aluminum. I've found a way to make the polish more cheaply by using a slightly different mix of ingredients and a revised heating procedure. What I get meets all the specs of our commercial product—etch rate, degree of flatness, residue, and the like."

"So what's the question?" queried Martin.

"Well, when you use this stuff on aluminum alloys, it behaves a little differently than what we sell now. The etch rate actually starts out a little higher, but the new stuff isn't as good at dissolving certain metals. So if you have aluminum mixed with a little copper, for example, the copper tends to build up on the surface. The etching still proceeds, but slower. Plus, you get a surface composition a little different from that of the bulk material."

"But I thought you said your new polish met our specs."

"It does, for any practical purpose I can think of," replied Conrad. "Usually in the applications this polish is used for you don't want to take off more than about 100 micrometers of material. Often it's much less. But because of the transition metal buildup, if you try to take more than 100 micrometers off, the etch rate eventually falls below our specification. And in any case, you get a surface more highly enriched in the resistant metal than you might expect. I can't see why that would ever matter, but conceivably it might."

"What are you asking me?" responded Martin.

"Since we meet our specs for practical applications, I want to try out this new formulation on some of our customers. But I can't see any up side to telling them what we're doing. Some customers get scared if you tell them you're tampering with their feed materials. And if you explain to them exactly what you're doing and what effect it might have, they don't always understand you. Some of our customers make toys! They don't know or care about etching chemistry! They just want something that will make their product look shiny the way it always has. They might get spooked and jump to another supplier. And we can't afford that right now. So I'm asking you if the change is OK."

Martin frowned. "I don't know, Conrad. What if something goes wrong? We think we know what our customers use this stuff for, but we don't know for sure. They may do other things we don't know about."

"I know there's a risk, but I think it's small," Conrad replied. "We can make a better profit with this new material, and it's hard for me to imagine an application where we wouldn't meet our specified etch rate. As for surface composition, we only warrant that our material doesn't deposit junk. We don't say anything about changing composition because of differential etching. So legally I think we're fine."

"I'm still not sure, Conrad. I'll have to think about it."

◆ What should Martin recommend?

CASE 14.4 Responsibility for Errors That Met Previous Safety Standards

"So that's how things are going at the plant, Monica. We're all wait-ing for you to get out of this hospital and come back to work. Do you think it'll be long now?" Emily Laborvincet finished her question with an expectant look.

"I think maybe another month," responded Monica Ichdien. "At least for part-time work. Actually, I hope to get out of the hospital in about two weeks. Then I'll be home for another two before think-ing seriously about work. But I'm definitely out of the woods. Luck-ily, I don't have to be in too much of a hurry, since you and the rest of the gang are doing so well without me!"

Emily grinned sheepishly. "It helped that everyone got a raise re-cently. Giving almost everyone the same 3-percent amount seemed to promote solidarity. And giving Conrad that little extra—3.5 percent—turned out to have the symbolic value we hoped. We just couldn't afford to lose our chief lab chemist now, and I was really worried he would leave. But he appreciated the special gesture." Monica nodded her approval. "And I wanted you to know," Emily continued, "that I never needed to help Tripos with that trade secret. I told you about it awhile back, remember? The improved procedure for making certain etchants that I learned at Beta Chemical last sum-mer."

"I'm glad," Monica responded. "Even though you didn't sign a non-disclosure agreement, using that procedure here at Tripos would've been too shady for my blood."

Emily sat thinking for a moment. Then suddenly she bolted up-right in her chair. "Oh, I just remembered. There's one more thing. During our renovation of the Tripos factory, one of the contractors happened to be sitting in our lunch room. He noticed that the paint is very old, and guessed that it contains lead. Sure enough—we had it tested, and it does. The problem is that in the past few months some of the workers have been bringing their children to lunch, pick-ing them up from the nearby grade school. The paint is peeling from the walls and ceiling, which is really bad with young kids around."

Monica nodded. "Right. We can't have that. Let's get it scraped and repainted."

Emily furrowed her brow. "Well, I checked into that, and it's not so easy. If we contract to an outside firm, the job is pretty expen-sive, mainly because they have to take all sorts of precautions to get all the paint off the walls and keep leaded dust from getting in the air. It'll run several thousand dollars. Even with the extra money we just got from your business partner, we don't have that kind of spare cash. We'll bust our building budget by about two grand. We can probably get away with it by some creative juggling of accounts, but it'll take some work."

Monica's eyes widened. "I didn't know putting a coat of paint on the walls was that expensive!"

"Usually it isn't, but now our lunch room is basically a toxic waste dump. Those are expensive to clean up."

"I really don't think scraping leaded paint is that big a deal for adults to do," Monica huffed. "They don't need to get every speck out. This is being overblown."

"I tried to tell that to the contractors I talked to," Emily responded. "But they said there are all kinds of liability issues when you do this kind of thing on a formal contract basis."

"Can't we just get some of our own floor workers to do it?" suggested Monica. "Either during work hours, or as overtime? That way it won't be so expensive."

"I thought about that too," Emily declared. "I asked around, and a few of our people are willing to do it. The trouble is, if they're not careful they might well get this leaded stuff all over the place. I don't know if there's any liability you have to worry about."

Monica looked perplexed. "I guess there's a remote chance that could happen," she muttered. "I guess we could always just ban children from the room, but I hate to do that. It also makes the adults feel a little unsafe. The bad thing is that this kind of paint met code with no problem when it was first applied. It's not like someone was being careless!"

"Well, you own the company. You have to tell me what to do," Emily answered.

◆ What should Monica recommend?

15

DEALING WITH DIFFERING ETHICAL SYSTEMS

"There are trivial truths and the great truths. The opposite of a trivial truth is plainly false. The opposite of a great truth is also true."
NIELS BOHR (1885–1962, NOBEL PRIZE—PHYSICS, 1922), CONTRIBUTION TO *THE NEW YORK TIMES*, OCTOBER 20, 1957

Even casual observation of how people act shows that many approaches to ethics exist. How do we choose among these approaches? Does it even matter what we choose? We will explore some aspects of these questions in the present chapter.

Differing Anthropologies

This book uses an anthropology, or model for the person, that sees the psyche as a unity of mind, emotions, and will. Other anthropologies exist as well, some having origins that are very ancient. Let's examine a few current anthropologies, and see what implications these have for ethical analysis.

Anthropologies based on modern psychology: A complete anthropology should account for the psychology of development, particularly as it affects moral behavior. Many theories exist to account for moral development.[1] Choosing among them affects the question of moral responsibility. For example, at what point do children become fully responsible for their actions? If there is a progression of responsibility along which children move, what description should we use, and how can we make the progress go faster? A complete anthropology should also account for disorders like psychosis, depression, and compulsion. Once again, no single comprehensive theory accounts for all, but whatever view we take affects the question of moral responsibility. For example, at what point does addiction destroy the responsibility of adults for their actions? If an employee suffers severely from addiction, our answer may determine whether that person is fired or merely required to seek professional help.

Anthropologies based on natural observation: Some anthropologies remain completely rooted at the level of what can be observed in the natural world. In this view (sometimes called positivism), people represent no more than the sum of their atoms and molecules, and disappear completely at death. Ethical behavior is then understood in terms of human pleasure, survival of the species, and the like. While avoiding problems with appeals to the supernatural, such an anthropology has problems justifying why people should do good in the face of unmerited suffering and uncertain rewards.

Anthropologies based on the supernatural: Some anthropologies appeal to things beyond the observable world. The most well known of these anthropologies originate in the world's long-standing religious traditions. Others include witchcraft and shamanism. All lay out in great detail the relation between humans and one or more supernatural beings, describe an immaterial dimension to human construction (e.g., a soul), and refer to a life after death. Looking beyond the observable world can fill in the gaps that plague anthropologies on the basis of nature alone. Unfortunately, differing supernatural anthropologies cannot be verified by systematic measurement, making difficult a choice among them and the moral systems they suggest. More serious problems arise when these anthropologies prescribe practices that harm human well-being—human sacrifice, for example.

Differing Principles and Methods

Separating principles from methods sometimes becomes a difficult task that we will not attempt here. Instead, we will summarize very briefly the highlights of major forms of ethical analysis.[2]

Egoism: Egoism represents more of a principle than a method of analysis. While egoism as a formal approach has found several defenders over the centuries,[3] we mention it here mainly because so many people use it in practice. Basically, an ethical egoist promotes his or her own good. Actions are considered right if they bring about more benefits than harms for the self. Notice that not all egoistic behavior has to be selfish; sincere concern for the welfare of others can lead to great benefits for the self in some cases.

Utilitarianism: Utilitarianism represents both a principle and a method (actually, several methods).[4] As a principle, utilitarianism appeals to the principle of utility: that actions should lead to consequences having the greatest total balance of benefits over harms. There are many utilitarian methods that vary according to how benefits and harms are defined and to whether the balance applies to individual acts or to rules that govern those acts.

Deontology (pronounced DEE-on-TOL-uh-jee): Instead of focusing on consequences the way egoism and utilitarianism do, deontology asserts that other features of an act determine whether it is right.[5] Deontological theories focus heavily on the general rules that govern duty or obligation, and demand that those rules be obeyed under all circumstances. These rules may come from divine command,[6] the state, or reasoned argument. Some forms of deontology (like "situational ethics") focus on what to do in individual cases, whereas others focus on general patterns of behavior to be followed in all cases.

Rights-based theories: Rights-based theories hold that people inherently possess certain rights, and that duties flow from the need to respect these rights.[7] The U.S. Declaration of Independence uses this idea in its famous reference to the endowment of people with inalienable rights to life, liberty, and the pursuit of happiness. Rights-based ideas underlie the approach of many groups who work on behalf of, for example, political prisoners, or the poor. Problems crop up in defining rights, however—some people tend to elevate their personal preferences to the status of inherent rights.

Intuitionism: Some people hold that many important elements of morality cannot be justified on rational grounds. These elements can be sensed and known only by direct experience or intuition, as in the way we experience the scent of a flower or the feelings of love. Some moral writers have pointed to a difference in the way men and women (on average) approach moral problems, and have suggested that women commonly frame their solutions with more attention to personal relationships than with a dry appeal to rational principles.[8] Some non-Western approaches to ethics do a similar thing.[9]

Casuistry (pronounced CAZH-oo-is-tree): Casuistry represents more of a method of analysis than a principle. Casuistry assesses an action by comparing it with two other "paradigm" actions, one that is clearly right and one that is clearly wrong. For example, suppose we want to judge whether it is right to kill a puppy for food. Paradigm actions might include killing off a carrot patch (right), and killing another person (wrong). Our decision depends on whether we classify the puppy as closer to a bunch of carrots or to a person. Casuistry's usefulness depends on whether we can come up with suitable paradigm actions. In our example, some might argue that carrots represent a poor paradigm when considering a puppy—some animal might be more appropriate. Of course, then we need to find a paradigm animal whose killing for food is clearly acceptable. Even with good paradigms, casuistry inevitably suffers from the problem of line-drawing. Casuistry casts every action as either right or wrong, with no allowance for degree. At the dividing line separating right from wrong, a tiny change in detail can shift the entire action over the border, counter to common sense.

Interestingly, among all the methods we have described here (other than virtue ethics), only intuitionism pays significant attention to interior morality. Egoism, utilitarianism, deontology, and rights-based ethics focus mainly on exterior actions, not intentions. Casuistry can incorporate intentions in principle, but in practice rarely finds use this way. The neglect of interior morality makes these approaches suspect. By accounting for interior morality, intuitionism at least in principle can serve as a useful basis for person-to-person morality. However, the vagueness of intuitionism makes it difficult to use as a tool for social policy, which requires concrete laws. Virtue ethics as described in this book seems to offer an advantage in this respect—the theory attends to interior morality while still offering a concrete framework for social policy.

Monism and Relativism

How should we handle all these different approaches to anthropology, principle, and method? Although they often conflict with each other, each seems to bring a perspective that contains an important germ of truth. Over the years people have responded in several ways to this problem.

Some have focused on similarities in the approaches, and argued that each perspective represents just one portion of a single, deeper ultimate reality. This view is roughly equivalent to monism—the belief that reality has only a single fundamental entity. Whatever truth this idea might hold, the tendency to gloss over real differences in practical moral rules remains a serious problem that can lead to a very superficial approach to ethical living. In situations where the various perspectives clash, the temptation to pick and choose expediently can become very strong, leading to a cafeteria-style form of ethics based on convenience.

Other people have argued that all approaches have equal validity. This view underlies relativism. It counts many defenders over the centuries, beginning with the Sophists of ancient Greece even before the time of Aristotle. Sometimes relativism arises out of a belief that humans can never learn what objective morality is, even if it exists in theory.[10] Other times this view comes from a belief that truth represents no more than a culturally-conditioned phenomenon with no objective validity.

One major danger of relativism has been known since the time of the Sophists. When carried to extremes, relativism can be used to justify ruthless, uncontrolled self-interest. Indeed, the Sophist Thrasymachus proclaimed "injustice pays," and went on to say that the appearance of justice serves only as a veil to protect the interests of the strong. He put the idea this way: "the sound conclusion is that what is 'right' is the same everywhere: the interest of the stronger party."[11] In other words, the practical consequence of a world where all forms of morality are created equal is that morality becomes enslaved by raw power.

Another major danger of relativism lies in its closeness to nihilism (pronounced NIGH-ul-izm or NEE-ul-izm)—the attitude that trying to be ethical makes no sense because there is no way to reconcile all the different approaches. In other words, the whole business of ethics becomes an exercise in futility. This attitude severely weakens the ability to commit to a serious ethical life. Indifference can follow as a logical consequence. Nihilism appears with enough frequency in modern society to make further examination worthwhile.

Postmodernism

Postmodernism represents a new mode of nihilistic thinking that has emerged on the cultural landscape over the past few decades.[12,13] Postmodernism does not speak directly to anthropology, principles, or methods in ethics, but includes a characteristic set of attitudes that has important implications for ethical behavior. Many writers still debate vigorously about exactly what postmodernism is. We will not argue the details here, but will note that our use of the term is looser and includes a less concretely defined social component than many formal philosophical treatments. In the view we take, core features of postmodernism include[14]:

1. a view of existence as lonely and impermanent, dominated by random happenings. Human feelings and efforts seem exhausted, and human values seem relative and arbitrary.

2. intellectual activity that refers to itself a lot, with big doses of absurdity, self-contradiction, and cynical satire. The bizarre and incoherent become commonplace.

3. a view that any kind of behavior is possible together with an attitude of coolness and detachment toward all things. Expressions of shock or surprise are suppressed and replaced with indifference or cynical laughter.

The postmodern view includes relativism as an important part. However, relativism by itself is normally an intellectual position that is accepted or rejected on the basis of rational thought. Postmodernism requires no such conscious choice. Just a little ordinary observation suggests that people can develop postmodern attitudes unconsciously through unpleasant life experience.

Postmodernism as we have described it poses dangers to the ethical life similar to those of relativism, but greatly sharpened by a sense of meaninglessness and hopelessness. These attitudes make sustained commitment to any form of morality very difficult. The rigors of everyday living pose major moral challenges even for the dedicated. Postmodern attitudes make these challenges worse.

True Pluralism

Our discussion argues that monism, relativism, and nihilism suffer from severe problems in the way they view the differing approaches to ethics. Yet ethical diversity remains an established fact that we must deal with. It probably makes sense to just accept this fact, but also to do our best to choose a good set of anthropology, principles, and methods. Having made this choice, we should try to adhere to it consistently, tweaking it and filling it out as experience suggests. Firm commitment in the face of difficulties plays a central role. Only after long consideration based on a pattern of failures should we consider giving up the core of whatever approach we have chosen in favor of another.

What about people who choose approaches differing from our own? How should they be handled? The analogy between ethical furniture making and moral living may prove helpful here. Both endeavors represent a craft. We have said before that several good ways may exist to build a cabinet, but some ways are better than others and some ways fail completely. Similarly, there may be several good approaches to carefully crafting a moral life. Experts in this craft do well to admire and learn from each other's actions, in the way that skilled furniture makers can admire and learn from each other's handiwork. There should be little question of trying to "convert" someone else from his or her basic point of view. Of course, just a little observation of our world shows that experts in the moral life are not very common. It makes sense for people who are not yet expert (but want to be) to find a master craftsperson and pay careful attention to what he or she does.

Conclusion

This chapter has offered no simple solutions to dealing with diversity in ethical approach. It has suggested that picking a particular approach and sticking to it offers the best hope for leading a consistently ethical life. The commitment requires the virtue of fortitude, while a periodic evaluation of how the approach is working requires prudence. Both efforts benefit greatly from the support of others, especially from those who clearly know what they are doing. When ethics is practiced as a craft—with avoidance of indifference, hopelessness, and sloppiness—pluralism does not have to be a bad thing.

A Real-life Case: Geological Experiments in Sacred Mountains

Seismographic data from earthquakes offer considerable information about the Earth's interior. However, the random occurrence of earthquakes in space and time limits their usefulness for this purpose. Some geologists attempt to overcome this problem by planting regular arrays of explosive charges under the ground in suitable areas and monitoring the vibrations that result from the controlled blasts. Such an experiment was planned in the early 1990s for the region around the Valles volcanic caldera in the Jemez Mountains of New Mexico. The so-called Jemez Tomography Experiment (JTEX) was intended to better understand the motion of magma deep within the Earth's crust, thereby permitting better assessment of the hazards posed by large volcanoes. The experiment involved several university and government laboratories, but experimental details were organized mainly by Los Alamos National Laboratory because it was near the site. Approvals for the experiments were required from the Department of Energy (which was responsible for Los Alamos) and the U.S. Forest Service, which controlled much of the land in the Jemez Mountains.

Problems arose when the Native American Pueblos in the nearby area learned of the experiment and lodged a formal protest with Los Alamos. While no explosion occurred directly on Pueblo-owned land, the Pueblos pointed to the sacred nature of the Jemez Mountains in their religious beliefs. Setting off explosives within these mountains was seen as perturbing the balance of nature, leading to possible unanticipated consequences for humankind. The Pueblos insisted that JTEX infringed on their sacred sites and practices. Problems multiplied because almost all aspects of Pueblo religion are secret, so that no specifics were offered to back the claims or to allow for negotiating a modified experiment. Los Alamos representatives got the distinct impression that they were missing the point by speaking in terms of specific sites or ceremonies—relatively Western concepts in this context. Given such limited information, the Forest Service decided to approve JTEX, reasoning that denial based on such vague claims would undermine the Service's ability to manage the nation's forests. However, wishing to avoid bad publicity on top of what it was already facing for handling the nations nuclear stockpile, the Department of Energy delayed its decision so long that Los Alamos decided to attempt a compromise.

After extended negotiations with the Pueblos, Los Alamos agreed to drop two high explosive blasts directly within the caldera, and replace them with vibreosis techniques. In vibreosis, a truck-mounted hydraulic apparatus vibrates the ground over a sweep of frequencies. The technique is more complicated than the setting off of explosives, and at the time its effectiveness for experiments like JTEX was unproven. Three of the four Pueblos in the region approved the compromise, and DOE likewise consented. The revised experiment took place successfully during the summer of 1993.

- ◆ Would you have approved the experiment as originally planned? Why or why not?
- ◆ In direct conflicts like this between scientific efforts and religious beliefs, which should take precedence if no compromise can be reached?

Reference

Baldridge, W. S., L. W. Braile, M. C. Fehler, and F. A. Moreno. "Science and Sociology Butt Heads in Tomography Experiment in Sacred Mountains." *EOS* 78 (1997):417–423.

"The uncommitted life isn't worth living."
MARSHALL W. FISHWICK (1923–)

Notes

1. For a good summary of these various developmental theories, see Daniel A. Helminiak, *Spiritual Development: An Interdisciplinary Study* (Chicago: Loyola University Press, 1987), ch. 3.

2. For a convenient but more detailed summary, see William Frankena, *Ethics*, 2nd ed. (Englewood Cliffs, N.J.: Prentice-Hall, 1973).

3. Famous writers who have defended ethical egoism include Epicurus (c. 342–270 B.C.), Thomas Hobbes (1588–1679), and Friedrich Nietzsche (1844–1900).

4. Famous historical utilitarians include Jeremy Bentham (1748–1832) and John Stuart Mill (1806–1873).

5. Famous deontologists include Socrates (c. 469–399 B.C.) and, with more precision, Immanuel Kant (1724–1804).

6. The Ten Commandments represent a good example of rules from divine command.

7. Originally put forth by John Locke (1632–1704), rights theories have become quite popular in the United States during the past few decades.

8. See Carol Gilligan, *In a Different Voice: Psychological Theory and Women's Development* (Cambridge, Mass.: Harvard University Press, 1982) for an excellent discussion of this gender difference.

9. For example, Hunter Havelin Adams in the *Portland Baseline Essay in Science* writes, "Nobody has a monopoly on truth. . . . There is no one correct way of knowing: there are ways of knowing. And Western conceptual methodology cannot discover any more basic truths to explain the mysteries of creation than can a symbolic/intuitive methodology." He continues, "For the ancient Egyptians, as well as contemporary Africans worldwide, there is no distinction between science and religion." (Quoted in Bernard Ortiz de Montellano, "Post-Modern Multiculturalism and Scientific Illiteracy," *APS News*, January 1998, 12). There is a group called the melanists who attribute a higher spirituality to people of color than to whites due to higher levels of melanin. The melanin supposedly has properties of superconductivity, high magnetic susceptibility, high absorption at all electromagnetic frequencies, and the capacity for information processing. Among other things, the melanin is claimed to offer greater potential for extrasensory abilities and deeper spirituality.

10. Indeed, the Sophist Protagoras (c. 480–c. 410 B.C.) took this agnostic view. Interestingly, he did not carry his relativism to the point of saying that every person should remain entirely free to act as he or she pleases. He argued that laws made by the state should be observed, not because they are the best possible, but because they are as good as can be made. See Samuel Enoch Stumpf, *Socrates to Sartre: A History of Philosophy*, 3rd ed. (New York: McGraw-Hill, 1982), 30–32.

11. Quoted in reference 10.

12. Specific examples of what some writers have classified as postmodern include films like Quentin Tarantino's *Pulp Fiction* and *Scream*, television programs like "Beavis and Butthead" and "Saturday Night Live," and rock music groups like Faith No More and Nine Inch Nails.

13. For a fascinating analysis of how the philosophical writings of even well-known physicists like Bohr, Born, Heisenberg, and Pauli have a closer relationship to postmodern criticisms of Western science than we might suppose, see Mara Beller, "The Sokal Hoax: At Whom Are We Laughing?" *Physics Today* (September, 1998):29–34.

14. This approach to postmodernism draws partly from Joseph F. Feeney, *America* 177 (November 15, 1997):12–16.

Problems

1. Write a page or two describing an ethical dilemma involving an ethical system different from yours that you have encountered in a job you've had. (If you've been lucky enough never to have been con-

fronted with a problem like this, describe one that a friend or relative of yours has had.) Recommend what action you think you (or your friend/relative) should have taken, and give reasons for and against that recommendation. Note: you don't have to say what was actually done in real life (unless you want to)!

2. Each case below has a question after it.

   a. List the options/suboptions available to the main character who has to make a decision, together with the event tree flowing from each option.

   b. Recommend what you think the character should do.

CASE 15.1 Dealing with Ethical Egoists

"What did you and Monica decide about painting that lunchroom at your factory?" Todd Cuibono asked his girlfriend Emily Laborvincet as they waited for the check at the Italian restaurant.

"She bit the bullet and decided to pay an outside contractor to get the old coat of leaded paint off," Emily responded. "And as accountant I had to be the one to figure out how to get the bill paid when we're stretched so thin financially."

Todd shook his head. "I still say she just should have had a couple of the factory workers do it. Sometimes I think she's not practical enough."

Emily rolled her eyes. "Yeah, I know. You're always so 'practical.' Anyway, how are things going at Pandarus Pizza? You haven't talked about it in a while."

"I expect they'll get better soon," Todd declared. Then he hushed his voice, glanced around apprehensively, and began to speak more earnestly. "Actually, Emily, I need to borrow something from you. You know that tape recorder you've got? The one with the attachment for recording telephone conversations? I need it for a few days."

"Whatever for?" Emily exclaimed quizzically. "What does that recorder have to do with Pandarus Pizza?"

"Shh!" Todd chided. "I've finally figured out how to get rid of the two people I hate most over there—my boss Thorne Mauvais and one of my subordinates, Celia Peccavi. With Thorne out of the way, I'll probably get his job! At least part-time, anyway. That restaurant doesn't need a full-time manager."

"What are you talking about?" Emily rasped incredulously.

"Well, lately Thorne and Celia have launched a total war on each other. They're at each other's throats all the time. Celia's been a snot toward him, and he's cut her hours to the bone. He's just looking for an excuse to fire her. Any way, each of them has accused the other of all sorts of things. Celia has accused Thorne of taking bribes from suppliers—a window painter I guess—and of permitting totally unsanitary practices in the restaurant. And Thorne has accused Celia of giving away free food when she's not supposed to. For both sides, I know some of this stuff is true, because I've seen it myself. But the rest of it I wasn't so sure about. Basically, they're at stalemate, because if one of them makes all these accusations to the owner, so will the other one, to retaliate."

"How does this involve you?" Emily persisted.

"Well, I decided to play like a mediator. In private I've tried to be really sympathetic to each of them. I've been doing it for about a month. Their hate is so strong that they're getting careless and letting their guard down. In fact, over the past week I basically got both of them to admit what the other has accused them of. In complete secrecy, of course. Now I just need to get it all on tape. If I can do that, I can give the tapes to the owner, and there's no question he'll just fire both of them!"

Emily stared in stunned silence for a moment. "I don't believe this!" she finally sputtered.

"What's not to believe?" Todd retorted with a hint of contempt. "They both deserve to be fired for what they did. And they're both jerks I can't stand. Plus I can get a better job by replacing Thorne. I can do the work a lot better than he can. No one's asking you to do anything. I just need your recorder. I mean, what's the world coming to if a guy can't borrow from his girlfriend?"

Emily just shook her head in disgust. "I don't know. It seems more and more that I just don't like the way you do things. It goes against my grain. Like the time there was that benzene spill where I work, and I had to decide whether to report it to the government. You told me not to, and I listened. But I never felt very good about it."

"It was good advice," Todd broke in. "Think of all the trouble you saved yourself and the company. And I do listen to your advice sometimes. Remember our lab partner Salina . . . how she didn't do anything on that big lab project, and we decided to write the final report ourselves without her? You insisted we complain to the professor, even though I thought it would do more harm than good. But we did it your way, even though we lost points in the process."

Emily scowled. "Yeah, about two—the way things turned out after she cut us a break on that crazy grading policy of hers. You were just taking the path of least resistance, Todd. You were willing to sacrifice a couple of points to keep me from bothering you about it. If the stakes had been higher, you would have fought me tooth and nail. About the recorder . . . I'll have to think about it."

◆ What should Emily do?

CASE 15.2 Working for a Company Whose Practices You Oppose

"Thanks again for watching the kids," said Dolores Sola wearily. "I get the shopping done faster when they're not around. Even just the two hours you give me every other Sunday helps a lot."

"No problem," replied Myra Weltschmerz as she put on her jacket. "They were good this time. Except when they were in the bathroom brushing their teeth. An argument started, and Garrett squeezed some toothpaste into Lorelei's hair."

Dolores rolled her eyes. "That's the second time this week! I can't get him to stop picking on his sister." Her voice hardened. "And it gets especially bad after they visit their father."

Myra nodded knowingly. "Uh-huh. I know. Remember, I came from a divorced family."

"He doesn't care to see them much," Dolores continued, "but sometimes they beg him and he gives in. Then he spends the whole weekend giving them candy and toys. There's no discipline at all. They tell him about all the times I don't give them what they want, and he eggs them on. Then the weekend ends, they come home, and I have to pick up the pieces. It's a nightmare!" Dolores paused, then added bitterly, "I want them to see their father, but sometimes it's more trouble than it's worth." Myra made a move toward the door. This nudged Dolores out of her sorrow for a moment. "Wait," she said hastily. "I'm sorry to dump this on you. Tell me about that interview you had last week. Was it good?"

Myra stopped and nodded. "Yeah. They offered me a plant trip!"

Dolores's face lit up. "That's *great!*" she enthused. She gave Myra a hug. "When are you going?"

Myra's face fell. "I'm not sure if I am."

Dolores joy changed to shock. "W-why?"

"Well, I didn't know much about the company before I interviewed. It's a chemical company that makes all kinds of things. They said they wanted an environmental engineer, which I am. But after the interview I happened to see this big article about the company in one of the trade journals I read. Just by luck. The article said they have a long record of pollution violations, with lots of fines. And it hasn't been improving. They've got all sorts of legal trouble over it that could bankrupt them if they lose enough cases. And they sell stuff overseas that's banned in this country because of toxicity or something. I'm not so sure I want to work at a place like that."

"But I thought you said the job market was tough right now. Do you have other trips lined up?

Myra shook her head. "No. That's the hard thing. My interviews aren't always that good. I don't seem to have enough confidence. But I didn't go to college for four years to make stuff that kills people!"

"Yeah, but take it from me, Myra. Unemployment is no good. I know, 'cause I've been there. You're always worried about whether you can pay the rent."

Myra nodded. "I know, but you have children. I don't have them yet. And I want to do something I like, that I'm proud of, you know?"

"Well maybe you can," Dolores contended, "even with this company! You said they wanted an environmental engineer. Maybe they're trying to clean up their act. If you can help change them from the inside, that's something to be proud of!"

"I don't know, Dolores. The article I read was pretty damning. There were lots of facts and figures, and interviews with former employees. I'm not a superhero. Why should I lead the charge against all those problems? My life is hard enough as it is. There are other ways for me to be proud and happy." Myra paused. "I don't want to visit them if I'm sure I don't want to work there. I don't think it's fair to them. And I haven't traveled much before, especially alone. It makes me nervous. So a plant trip is a lot of stress for me."

Dolores gave Myra another hug. "Did you talk to your boyfriend about it? You know, Martin?"

Myra nodded. "Uh, huh. But he just told me to do what I think is best." Then she turned to leave. "I've got to go, Dolores."

◆ What should Myra do?

CASE 15.3 Research on Animals

Terence Nonliquet breathed a sigh of relief as he erased the blackboard. He enjoyed his job as a teaching assistant in computer science at NTI, but he was glad the semester had finally ended with today's class. Only the final exam remained. When he finished, he banged the eraser down emphatically with a sense of relieved enthusiasm. However, his stomach sank when he turned around to see, not an empty classroom, but one final student to deal with—Celia Peccavi. Terence raised his eyebrows. "What do *you* want?"

Celia kept her distance. "Now that the semester is over, I just wanted to apologize for . . . well . . . you know . . . a couple of days ago."

"I can think of a lot of things you could apologize for," Terence retorted. "Like cheating and plagiarizing on your homework. Like repeatedly trying to extort me into a date with you. And like trying to drive a wedge between me and my girlfriend."

Celia reddened slightly with embarrassment. "Yeah, I pushed too hard. I'm sorry."

Terence eyed her suspiciously. "That's all you want?"

"Well, I have two other things. I wanted to thank you for normalizing the grades between sections the way you did. You know . . . including the homework as well as the exams. I heard you and the other TA's secretly went against Professor Bligh's instructions. I got a C instead of a D because of it."

Terence frowned slightly. "It's not much of a secret if you know about it. I know you overheard me talking on the phone about the plan that one evening, but how did you find out what we eventually did?"

"I overheard one of the other TA's talking about it later," Celia answered. "Don't worry, I won't say anything. I just wanted to thank you."

"I did it because it was fair, not because it helped you. What was the other thing you wanted?"

"I know," Celia responded hastily. She paused, then ventured, "I wanted to ask your advice, too." Terence crossed his arms silently. "You know, I got fired from Pandarus Pizza," she continued.

"No, I didn't know," Terence replied without expression.

"It was a big mess, very unfair," she said sadly. She eyed him for signs of sympathy, but saw none. Her voice became more businesslike. "Anyway, I looked around, and got an offer from a different company to work over the summer. Actually, it's two offers since they're giving me a choice of two jobs. It's a firm that tests personal care products on animals. You know, shampoos, cosmetics, that sort of thing. They look mostly for skin irritation, using mainly rabbits. One of the jobs involves just entering the data some lab person takes into a computer, collating the results, doing some simple statistics, and that sort of thing. There's no direct contact with the animals, and the pay is pretty good. The other job involves actually helping with the experiments in the lab. I'd have to care for the animals, monitor what happens to them in the experiments, and sometimes even help with applying the test materials onto the skin. That job pays great— 50 percent more, and I can really use the money.

"So why are you telling this to me?" Terence broke in somewhat impatiently.

"Well, I know you pretty well, and I respect your judgment. I'd like your advice on what to do."

"I've heard this from you before," responded Terence guardedly. "Last semester you wanted advice on how to handle some crazy situation at Pandarus Pizza. It was just an excuse to wriggle your way into my life."

"No it wasn't!" Celia protested. "I admit I sort of liked you, but I was doing badly in class and you asked me why. So I told you, and asked your advice. I really wanted to know." Terence frowned, and Celia continued in earnest. "Terence, I don't get along with my family well enough to ask them. I can't ask anyone I used to know at

Pandarus because I don't see them any more. And most of my friends aren't too mature. There aren't a lot of people I can ask. I swear I'm not trying to push you into anything."

Terence's expression softened slightly. "Well, I don't know. . . ."

"I mean, it's not like I wouldn't go out with you if you asked," she ventured tentatively, eyeing him closely. "I did hear you and Leah aren't together any more. But I definitely wouldn't try to force anything. I found out that doesn't work with you."

Terence stiffened at the reference to his former girlfriend. "Yeah, I suppose you must be happy. You were trying to get me away from Leah all along. But it wasn't just your antics that made us break up. There were all sorts of things. She and I finally agreed to just move on, that we weren't as compatible as we thought. I'm not looking for anyone else right now." He eyeballed Celia. "But even if I were, it wouldn't be you!"

Sensing defeat, Celia returned to her main objective. "OK, fine. I understand. But I still want to know what you think. I need the money to pay my tuition next fall. But putting slimy stuff on bunnies and watching them get a rash isn't my idea of fun. I'm not even sure if it's right. What do you think?"

Terence drew a deep breath, wishing for the conversation to end. "I don't know, Celia. Different people have different ideas on that. I can't make that decision for you." He gathered his papers and turned to leave. "If you don't mind, I've got a bunch of stuff to do this afternoon. I'm supposed to meet with Professor Bligh in ten minutes. Is there anything else?"

"No, that's OK," she responded quickly, and motioned for him to go.

◆ Which, if either, of the jobs should Celia take?

CASE 15.4 The Goals of Western Science

In response to a loud knock, Leah Nonlibet opened the door to the geology laboratory at NTI where she worked as an undergraduate research assistant. "Come on in, Brenda! Did you find the place OK?"

"Uh, huh," Brenda replied, looking intently around the lab. "It was nice of you to invite me here. We Poly Sci majors don't get much chance to see the inside of a research laboratory! So remind me. How long did you say you've been working here?"

"About a year and a half," responded Leah. "I mostly help the graduate students with their experiments and keep the lab drawers stocked, but this semester I've been doing some experiments on my own, too." Leah moved back toward the sink where she had been cleaning glassware. "If you don't mind, I just want to finish cleaning this stuff. You're a little early. I'll give you the nickel tour in a minute when I'm done."

"No problem," agreed Brenda, eyeing what Leah was doing. "You invited me here partly because you're not totally happy with what you do here, right?"

"Yeah. It's like I was telling you last week. My boss, Professor Clark, wants me to work here again next year when I'm a senior. But I'm not sure I want to. I don't like some of what goes on around here, and I'm trying to decide what to do. I wanted advice from a different perspective. You and I have been friends now for a couple of years."

"Uh, huh," Brenda laughed. "Since freshman chemistry! And I can see right away one thing I don't like. You're doing the dishes! You kill yourself studying in all these classes, but what you get paid to do is dishes! How many women spend their whole lives doing dishes? Can't you do something else?"

"They have to get done, by me or someone else," replied Leah. "It's part of the job description, whether I hold the job or a man does."

"How can you be sure?" Brenda retorted suspiciously. "Lots of job descriptions are tailored around specific people. So anyway, remind me again about all the stuff you had to fight about this semester."

"Well, first Professor Clark and I were going to publish a paper in a journal, and he wanted to leave out some key details of a proce-dure we developed. He said his competitors would rush to do some obvious follow-up experiments that our work suggested, and would leave us in the dust because we don't have the resources they do. Withholding some of the details for a little while would help us stay competitive. Then he wanted to leave some data points I took off a graph we were going to publish because he said they didn't make sense. Finally, he was thinking about making me second author on a paper where I took most of the data, and I thought I should be first author."

Brenda shook her head in disbelief. "I can't believe you'd stay here another minute, let along think about working here next se-mester!"

"It's not quite as bad as I made it sound," protested Leah. "I guess reasonable people could disagree about the data points, at least. Maybe about the other things, too. I didn't tell you all the details just now."

"Still," contended Brenda, "it's so obvious. The whole effort here is about power, prestige, and money. It's the patriarchal power struc-ture all over again. Clark and his henchmen don't worry so much about what they find as about how much power and recognition they'll get."

Leah thought for a moment. "I don't know, maybe there's some-thing to what you say," she began slowly. "It's true that Professor Clark rejected a journal manuscript that came from a small, out-of-the-way country because he said the paper wouldn't be interesting to most of the readership—in the United States and Europe. The pa-per was about some data analysis method for use with a hand cal-culator. Most labs do the analysis on a PC, but this article was meant for people in places where PC's are rare and expensive."

"Obvious bias," Brenda rejoined.

"But it was more complicated than that. The paper wasn't written very well. Professor Clark understood exactly what value judgment he had to make." Leah paused, then continued, "I know it's an im-perfect lab, but it's an imperfect world, too. And I like the science we do. We learn real truths about the way nature works. I helped find a new crystalline phase of a mineral, for example. It was really exciting!"

"What do you mean, 'truth'?" Brenda sniffed. "There is no truth outside of culture. It's all a conditioned thing. You can't understand any statement of 'truth' outside the culture in which it was said. For example, who cares about this new 'crystalline phase' aside from Clark and a bunch of other men. Do you think if Clark, his gradu-ate students, and all other geologists were women that you would even be studying this? You'd be asking totally different questions. You'd be looking at how minerals could be used for world peace, or better child care, or stuff like that. If you work in this lab, all you're doing is letting yourself be used like a pawn in a male-dominated power structure."

"Are you telling me I shouldn't work in any geology lab at all?" asked Leah incredulously.

"Maybe. If all they do is study things decided on by men. I don't know too many geologists who are women. I think your job is to protest, or at least not get contaminated by that kind of thinking."

Leah shook her head. "I like the subject. Even Professor Clark is pretty nice, although he and the other graduate students have faults. And while you look down your nose at money, I do need some. There aren't a lot of other labs in this department where I can go. They're not looking for undergrads like me right now."

"Well, I think you should quit. That's my advice," responded Brenda.

◆ Should Leah continue to work in Professor Clark's laboratory?

16

HABIT AND INTUITION

"The heart has its reasons which reason does not understand."

BLAISE PASCAL (1623–1662), *PENSEES*, NO. 423

Many methods for ethical decision-making focus on the thought process for making judgments. In this chapter we explore in more detail the advantages and limitations of this perspective, and attempt to sketch the outlines of a more comprehensive view.

Rationalist Approaches to Moral Action

Many models for moral action base themselves mainly on the workings of the mind and will, largely ignoring the emotions. These "rationalist" models take highly varied forms, and describing all of them lies beyond the scope of this book. Nevertheless, at the risk of much oversimplification and incompleteness, we list a few of the more common ones and highlight their main features.

Static models: Some models do not speak to questions of moral development, but instead focus on adults that are fully developed as moral creatures. These models are called "static," and are often used by philosophers. We described many of these models concisely in the last chapter: deontology, utilitarianism, rights-based theories, and casuistry. For brevity, we will not reproduce those descriptions here.

Developmental models: Other models speak directly to questions of development. Most focus on moral development in children, but some also address the changes that take place throughout adult life. Many of these models treat areas of human psychic growth that include more than just the moral, although the moral is certainly included. For example, Eric Erickson lays out eight "psycho-social" stages.[1,2] Each stage involves the resolution of a type of psychological tension, like trust versus mistrust, intimacy versus isolation, and social responsibility versus stagnation. The psychologist Jean Piaget, on the other hand, defines five stages of cognitive development with names like "sensorimotor," "concrete operational," and "post-formal operational."[3] Since cognitive function clearly enters into moral decision-making, such a theory speaks at least indirectly to ethics.[4] Some of these models, like that of Erickson, assume that stages of growth arise out of the natural process of maturation, and therefore depend mostly on age. Roger Gould[5] and Daniel Levinson[6] have also proposed such models. Other approaches question this view, and suggest that moral development can slow or stop in a way largely separated from a person's age. Piaget's model along with those of Lawrence Kohlberg,[7] James Fowler,[8] and Jean Loevinger,[9] follow these lines. Among all these models, Kohlberg's concerns itself most directly with moral growth. Kohlberg's original theory includes six stages, ending where a person acts according to universal rational principles.

Advantages of Rationalist Approaches

The approaches we have just described carry significant advantages, both for day-to-day ethical decision-making and for larger, more complex questions. These advantages include:

Comprehensiveness and completeness: An ethical approach with firm grounding in the mind offers the surest defense against rash decisions. Prudence helps guard against snap judgments by reasoning through all the possible options. This process guards against incompleteness or inconsistency in judgment, particularly when emotions run high.

Ease of use in discussion and debate: A rationalist approach permits direct recognition of what leads to an ethical position. The specific reasons people disagree become readily apparent—a great advantage in a pluralistic society where disagreements about ethics abound. These disagreements often demand compromises, which can be made more easily when everyone understands exactly where the compromises must be made.

Ease of use for legislation: Large, pluralistic societies require laws and regulations to maintain social order. In a representative democracy, these laws arise through a formal process involving several branches of elected government. The process of writing laws about ethics helps ensure that the underlying reasoning is laid out openly in a way that everyone can understand.

Problems with Rationalist Approaches

As significant as these advantages are, a purely rationalistic approach also carries several significant disadvantages. These include:

Inability to handle truly complex cases: Moralists going back at least to the Middle Ages[10] have understood that rational analysis has limitations when dealing with truly complex cases. Problems with concretely describing many of the abstract, fuzzy features of complicated situations plague all rationalist approaches to ethics. Hence, comprehensive theories for handling ethical complexity are rare and often unconvincing.

Sterility: Many rationalist theories come across as dry, abstract, and distant—far removed from the concerns of real people. This problem runs deeper than the inability to inspire or entertain (as important as that might be). Such theories may actually lead to defective moral conclusions. For example, Carol Gilligan has roundly criticized Lawrence Kohlberg's theory on these grounds.[11] She argues that women often "underperform" men in sociological studies of morality because the tests are written in terms of abstract duties and obligations. Gilligan contends that many women respond to values of community and personal commitment that are equally lofty as the values of men but are less easily expressed as abstract principles.

Excessive reliance on knowledge: Many rationalist theories rely upon an idea that was introduced by the early Greeks and built upon by other philosophers through the ages[12]: that people will naturally do the ethical thing if they know what it is. In other words, ethical failures come largely out of ignorance—ignorance of the future, of general principles, of the details of a particular case, and so on. Unfortunately, this idea flies in the face of much human experience. We can list a few counterexamples:

ADDICTION: Many people try hard to give up addictions to food, drugs, alcohol, sex, cigarettes, and the like because their behavior hurts themselves and others. Yet even when armed with buckets full of good reasons and oceans of determination, many of them fail. It's difficult to explain these failures by saying that the addicts do not know what they are doing or how to escape.

THE MENTALLY CHALLENGED: There is no convincing evidence that those whose mental functioning is impaired act less ethically than the rest of the population. Indeed, if morality correlates with intelligence, we would expect smarter people to be more ethical. Common experience does not support this idea.

THE PSYCHOLOGICALLY DAMAGED: Modern psychology shows that many people who have traumatic experiences early in life have an increased tendency for impulsivity, lying, stealing, violence, verbal abuse, and the like. Such behavior often seems to spring from a combination of fear and low self-esteem. Rooted deep within the unconscious, these tendencies apparently have little to do with conscious knowledge or reasoning.

SOCIOPATHS: Certain rare individuals act with no regard for the welfare of others. Classic examples include those who murder with no remorse whatever. Most of these people show no defect in knowledge or mental functioning.

MYSTICS AND NON-WESTERN ETHICS: History abounds with mystics, poets, monks, and others both in the West and elsewhere who appeal to the symbolic or mystical as the basis for relating to others. Such people often live heroically moral lives, apparently without resorting to complex reasoning. Indeed, some non-Western religions largely reject rational approaches to living more deeply, as with the "koans" or riddles posed by Buddhist masters to their students.

Failure to account for nonrational aids to ethical behavior: Many people pursue the ethical life using support groups, religious ceremonies, Twelve Step programs, prayers, counselors, spiritual directors, and a host of other people and activities. While some of these aids do appeal to the mind, many seek to inspire the emotions. Rationalist theories do not account for (or sometimes even consider) the obvious and enduring power of these nonrational aids.

Failure to offer a complete justification for ethical behavior: Rationalist approaches generally fail to answer completely a question of profound importance: why be ethical? For example, we have noted that Lawrence Kohlberg's theory of moral development ends when a person thinks in terms of universal rational principles. But why should anyone's life be ordered according to such principles?

In an attempt to solve this problem, Kohlberg has added an extra "Stage 7" to his theory that is explicitly spiritual or religious.[13]

The danger of rationalization: We have given many reasons why ethical behavior cannot be governed by rationality alone. Great danger lurks in an ethical world view that ignores the irrational elements of human action. Rationalization probably tops the list of moral failures that can result. That is, the irrational jealousy, envy, bias, and malice that smolder within virtually every human being creep in unnoticed to skew moral analysis, sometimes enough to make it surreal. Scientists and engineers, who normally work in an environment that highly values rational thought, can be particularly vulnerable to this kind of thinking.

Toward a More Comprehensive Approach to Moral Behavior

The disadvantages to rationalistic ethics have recently inspired attempts to fill in the gaps.[14] For example, the psychologists James Rest and coworkers[15] have introduced a "four-component model" for moral decision-making that is particularly simple to understand. This model focuses both on individual actions and on general orientation of life. On the level of individual actions, the model proposes four steps to moral decision-making: sensing the presence of moral issues, reasoning through them, making a decision, and following through on the decision. On the level of general orientation, the model implicitly appeals to four corresponding items: sensing, reasoning, judging, and doing. Thus, the four-component model resembles classical virtue theory in some ways. After all, prudence underlies sensitivity and reasoning, justice underlies judging, and temperance/fortitude underlies doing. Rest's model allows for "affect" (in our terminology, emotions) to influence decision-making in addition to the mind.

In philosophy, virtue theory is flowering again after lying dormant for several decades.[16–20] As we have seen throughout this book, virtue theory employs the idea of habit to place individual acts in the context of a broader orientation of life.[21,22] Contemporary virtue theorists argue that this approach represents an advance because the goal of moral life focuses on development of habits that form "character." This character offers the power to do what is good easily and without a great deal of reflection.

A REAL-LIFE CASE: The Ethics of Human Cloning

Few scientific events have created a firestorm of ethical debate like the cloning of the sheep "Dolly" in 1996. Although the cloning method used was difficult and inefficient, it seemed clear to many that the cloning of humans had moved much closer into the realm of genuine possibility. Reaction to the idea of human cloning was instant and mostly negative. Indeed, Nobel Prize winner Joseph Rotblat condemned the endeavor as "out of control," creating "a means of mass destruction," while the respected German newspaper *Die Welt* wrote, "the cloning of human beings would fit precisely into Adoph Hitler's worldview." In the United States, President Clinton imposed a five-year ban on federally funded human cloning research.

The issues surrounding human cloning remain quite complex. More than one technique exists. "Dolly" was cloned by "nuclear substitution," in which a nucleus from a developed, differentiated adult cell was placed into an egg cell containing no nucleus and then implanted into the uterus of an adult female. This procedure has not been used successfully with humans. However, human cloning of a sort has already been accomplished by "embryo splitting," in which the physical division of a very early-stage embryo provides twins: one for biopsy and the other for implantation. In both cases the success rate is not very high, resulting in a significant number of destroyed embryos. However, embryo splitting also involves the deliberate destruction of an otherwise viable embryo, whereas nuclear substitution does not. Should cloning techniques improve to reduce the number of failures, new issues arise including unforeseen genetic damage in the clones, the social status of the clones, attempts toward eugenics, and potential decreases in genetic variability.

Some of the most important issues seem unsolvable at the purely rational level, however. Many people believe humans have a crucial immaterial aspect—a soul. In this view, cloning's rather mechanistic approach to human life offends deeply against human dignity. Notions like "soul" and "dignity" cannot be approached through reason alone. Other people see cloning as the distillation of the persistent human tendency to attempt to control and use others. Here, the issue is motive—another notoriously difficult subject.

◆ What other issues can you name that surround human cloning?

◆ Do you think human cloning should be allowed? Under what conditions?

References

Benatar, D. "Cloning and Ethics." *Monthly Journal of the Association of Physicians* 91 (1998):165–166.

Katcher, F. "Cloning, Dignity and Ethical Revisionism." *Nature* 388 (1997):320.

Shapiro, H. T. "Ethical and Policy Issues of Human Cloning." *Science* 277 (1997):195–196.

"Most powerful is he who has himself in his own power."

LUCIUS ANNAEUS SENECA (C. 4 B.C.–A.D. 65), *EPISTOLAE MORALIS*

Notes

1. Erik Erickson, "Eight Ages of Man," in *Childhood and Society*, 2nd ed. (New York: W.W. Norton, 1963).

2. For an incisive summary of these various developmental theories, see Daniel A Helminiak, *Spiritual Development: An Interdisciplinary Study* (Chicago: Loyola University Press, 1987), ch. 3.

3. Jean Piaget, *The Origins of Intelligence in Children* (New York: Norton Library, 1963, originally published in 1936).

4. Jean Piaget, *The Moral Judgment of the Child* (New York: The Free Press, 1965, originally published in 1929).

5. Roger L. Gould, *Transformations: Growth and Change in Adult Life* (New York: Simon and Schuster, 1978).

6. Daniel Levinson, *The Seasons of a Man's Life* (New York: Knopf, 1978).

7. Lawrence Kohlberg, "The Child as Moral Philosopher," *Psychology Today*, September 1968, 25–30; "The Implications of Moral Stages for Adult Education," *Religious Education* 72 (1977):183–201; "Stages and Sequence: The Cognitive-Developmental Approach to Socialization," in *Handbook of Socialization Theory and Research*, D. A. Goslin, ed. (Chicago: Rand McNally, 1969).

8. James Fowler, *Stages of Faith: The Psychology of Human Development and the Quest for Meaning* (San Francisco: Harper & Row, 1981).

9. Jane Loevinger, *Ego Development* (San Francisco: Jossey-Bass, 1977).

10. Thomas Aquinas (c. 1225–1274), the leading philosopher of the Middle Ages in the West, argued that people have a "natural judgment" concerning certain moral goals like life, truth, and fairness. Aquinas's "natural judgment" clearly refers to a power to distinguish good from bad by means other than pure reason—that is, by intuition. See Thomas Aquinas, *Summa Theologica* (Westminster, Md.: Christian Classics, 1981), Question 47, Article 15.

11. Carol Gilligan, *In a Different Voice: Psychological Theory and Women's Development* (Cambridge, Mass.: Harvard University Press, 1982).

12. Immanuel Kant (1724–1804), most notably. Since the influence of Kant on current ethical theory is large, this premise pervades much talk about ethics in Western society.

13. Lawrence Kohlberg and Clark Power, "Moral Development, Religious Thinking, and the Question of a Seventh Stage," *The Philosophy of Moral Development: Moral Stages and the Idea of Justice, Essays on Moral Development*, Vol. 1 (San Francisco: Harper & Row, 1981), 311–372.

14. Although this book focuses specifically on ethics in science and engineering, we note in passing that some writers have attempted to set the whole endeavor of science and engineering in a more complete (and less rationalistic) context. Some of these writers are scien-

tists themselves. For just a few examples of this extensive literature, see Michael Polanyi (a physical chemist), *Personal Knowledge* (Chicago: University of Chicago Press, 1974); Stanley L. Jaki (a physicist), *Is There a Universe?* (New York: Wethersfield Institute, 1993); and Wolfgang Smith (a mathematician/physicist), *Cosmos and Transcendence* (Peru, Ill.: Sherwood Sugden, 1984).

15. James Rest, Muriel Bebeau, and Joseph Volker, "An Overview of the Psychology of Morality," in *Moral Development: Advances in Research and Theory*, James Rest, ed. (Westport, Conn.: Greenwood Publishing, 1986), 1–27.

16. Russell Hittinger, *A Critique of the New Natural Law Theory* (South Bend, Ind.: University of Notre Dame Press, 1987).

17. Alisdair MacIntyre, *After Virtue: A Study in Moral Theory* (South Bend, Ind.: University of Notre Dame Press, 1980).

18. Gilbert Meilaender, *The Theory and Practice of Virtue* (South Bend, Ind.: University of Notre Dame Press, 1984).

19. Romanus Cessario, *The Moral Virtues and Theological Ethics* (South Bend, Ind.: University of Notre Dame Press, 1991).

20. James D. Wallace, *Virtues and Vices* (Ithaca, N.Y.: Cornell University Press, 1978)

21. As we have said throughout this book, such an approach is not new. It was originated by Aristotle in ancient Greece, and further developed during succeeding centuries. For example, Thomas Aquinas in the Middle Ages speaks of "acquired prudence" that comes from time and experience. He indicates this "acquired prudence is what I think is the specific virtue that enables the elderly and experienced to adjudicate complex situations." See Aquinas, *Summa Theologica*, Question 47, Article 14, Objection 3.

22. It is curious just how strong the lure of purely rational approaches can be. While appealing to underpinnings of habit, many writings nominally based in virtue theory still have trouble making a practical and specific connection between habits and particular acts. Often the analyses wind up with a very act-oriented perspective after all, especially when dealing with questions of justice.

SUMMARY

This unit has examined several advanced issues in moral thought. If this focus seems out of step with the word "fundamental" in the title of the book, recall that "fundamental" means "foundational" or "primary," not necessarily "elementary" or "simple." In fact, important aspects of ethics that face most scientists and engineers over the course of their careers are neither simple nor elementary.

In Chapter 13 we discussed methods for allocating resources. We listed several methods and showed that no simple rules can guarantee fairness under all circumstances. The ultimate decision depends heavily on what needs to be distributed and on the specific details of each situation. However, sometimes we can obtain guidance from the principle that the obligation to avoid what is bad outweighs the obligation to do what is good.

In Chapter 14 we examined the concept of risk. We saw how scientists and engineers as well as laypeople balance benefits against harms when assessing risk. However, in laypeople the process involves many factors apart from a formal risk-benefit analysis. These include the degree of control they have in choosing, the way information is presented, the aversion to harm versus the desire for benefit, and proximity to the danger. We saw how making decisions about risk involves answering two difficult questions: what criteria should be used, and who should make the decision? For the latter question, issues of how expert opinion should enter in become paramount.

In Chapter 15 we looked at the difficulty of acting morally in the face of ethical systems that differ from our own. We sketched a few common but widely varying anthropologies, principles, and methods. This diversity can induce a lapse into ethical monism, which holds that the differences are really only superficial and obscure a deeper unity. Monism can easily slide into a cafeteria-style, pick-and-choose ethics based on expedience. On the other hand, the diversity of approaches can inspire ethical relativism, which holds that the different perspectives all have equal validity so that it does not matter which one is chosen. When pursued to its logical conclusion, relativism can slide into uncontrolled self-interest or into nihilism. In recent years, this nihilism has manifested itself as postmodernism—a body of attitudes characterized by loneliness, meaninglessness, self-absorption, and ironic incoherence. We suggested that the best way to handle ethical diversity is to choose a particular anthropology and set of principles and methods. Once this choice is made, we should adhere consistently to it, filling it out as experience suggests. Only after long consideration based on a pattern of failures should we consider giving up the core of the approach we have chosen in favor of another.

Finally, in Chapter 16 we have examined the diversity of ethical approaches from a different angle focusing on the limitations of rationality. We saw how rationalist perspectives carry advantages for complete, coherent debate and for social policy, but fail to account for many aspects of good moral behavior.

Some Words of Caution

We have now come to the end of our study of ethics. This may be the first book you have ever read on the subject. Hopefully the ideas we have outlined make sense. Hopefully you learned something useful. If so, this book has accomplished its purpose.

However, it's entirely possible that you have found ideas not to your liking. Some aspects may have been overemphasized, some underemphasized, and some ignored completely. The complex issues we have raised, particularly later in the book, have posed more questions than they have answered. This may leave a sense of incompleteness or dissatisfaction. Then the real danger looms that the study of ethics may seem useless—a tired exercise in futility, good only for show. If the study of ethics goes this way, the practice may follow close behind. However, if this danger is avoided, the book may yet have accomplished its purpose. If now you better understand where ethical issues in science and engineering lie, and if you have developed a more coherent way of thinking about them (regardless of whether you use this book's method), that counts as a success.

CASES: THE REST OF THE STORY

Martin Diesirae and **Myra Weltschmerz** both graduated at the end of the semester. With regard to the APZ aluminum polish (Case 14.3), Martin and Conrad finally agreed to say nothing to their customers about the reformulation. No problems materialized. Myra did decide to take a plant trip to the company with the bad environmental record (Case 15.2). However, she interviewed poorly there and was not offered a job. Martin and Myra stayed the summer in Exodus, he working for Tripos Metal Polish full time and she baby-sitting for Dolores Sola and working other odd jobs. By the following September they both found jobs appropriate to their degrees in the Chicago area. They moved there, and got married a year later. They had twins shortly thereafter, and Myra quit her job to stay at home until the children began school. Martin never really shed his rough edges, and Myra never gained full self-confidence. They drifted in and out of marriage counseling, and separated briefly, but never divorced.

For **Emily Laborvincet,** the tape-recorder incident with **Todd Cuibono** turned out to be the last straw (Case 15.1). She refused to loan him the recorder, and dumped him for good measure. She completed her chemistry degree and went on to get an MBA. She worked as an international business consultant for several years, marrying someone at her firm. Shocked by the breakup with Emily and unexpectedly foiled in his plan to get his coworkers fired, Todd quit Pandarus Pizza. He completed his degree in chemical engineering, found a job with a big company, and moved rapidly through the managerial ranks, ultimately becoming senior vice-president. He waded through great self-inflicted difficulties in his relationships outside work, however.

Terence Nonliquet finished his degree in electrical engineering and went on to graduate school in the same field. He eventually became a professor in that field in a teaching (not research) university, well liked by his students. He did not marry until his late 30s, but lived happily with his stay-at-home wife and three children thereafter.

Leah Nonlibet decided to keep her job in Professor Clark's laboratory (Case 15.4). A year later she graduated in geology and found a job with a small firm that contracts to companies that drill for oil. She ultimately became a successful mid-level manager. She had several lengthy relationships, but never married. Throughout her life she retained a vague sense of discontentedness, but could never really act to dispel it.

Celia Peccavi took the part-time job putting consumer products on rabbits (Case 15.3), but in school drifted from major to major. She eventually graduated after five and a half years with a degree in finance, and she found a job with a local bank in Exodus. Her personal life lurched from one swirling tempest to another.

INDEX

Ability to pay, allocation of resources by, 209
Accreditation Board for Engineering and
    Technology (ABET), xiii
Actions,
    as affected by circumstances, 38–40
    analyzed by exterior consequences,
      35–43, 88–95
    analyzed by interior consequences, 36,
      52–58, 107–10
    as ends, 41
    intention of, 52, 67
    as means to ends, 40–41
Adams, Hunter Havelin, 244n.9
Addiction, 255–56
Adeny, Douglas, 150n.5
Affirmative action, 212–13
Agency model for the professional-client
    relationship, 178
Allocation of resources. *See* Resources,
    allocation of
American Medical Assocation, 43
Analogy for ethics, mathematical, 88–89, 107
Animal testing, 78–79
Anthropology
    choice of, and ethical analysis, 20, 236–37
    components of ancient Greek, 20
Approval. *See* Intention
Aquinas, Thomas, 150n.1, 259n.10, 260n.21
Aristotle, 1, 8, 52, 180, 240
Augustine of Hippo, 80n.5
Authorship, order of in publication, 175–76

Babage, Charles, 160
Bacon, Francis, 173
Balancing
    consequences, 36–42, 88–95, 106–10
    interior and exterior goodness, 110–11
    use of likelihoods in, 89–90
Baltimore, David, xiii
Basil the Great, 205
Bellar, Mara, 244n.13
Benevolence, 56–57, 78, 161
    defined, 56
Bentham, Jeremy, 244n.4

Bhopal, 96–97
Bohr, Niels, 236
Breast implants, 42
Bruno, Giordano, 11
Butler, Samuel, 203

Caesar, Caius Julius, 228
Canute the Great, 193
Cardano, 228n.1
Casuistry, 238–39
Censorship of the Internet, 148–49
Challenger spaceship, xiii, 10–11
Character, 10, 257
Childress, James, 228n.2
Choice
    of attitudes, 107
    of principles and methods, 71
    relation to habit, 23–24
Chrysostom, John, 106
Cicero, Marcus Tullius, 3, 88
Circumstances
    effects on moral analysis, 38–40
    method of enumeration, 39–40
Clients, professional, 178–79
Cloning, 257–58
Codes of professional ethics, 9–10
Cold fusion, 190
Common good, 21
Communications Decency Act, 149
Community
    and ethics, 21
    and scientific knowledge, 158
Comparative negligence, 126
Complexity, ethical, xv, 71, 88–90, 254–55,
    259n.10
Comprehensive Environmental Response,
    Compensation and Liability Act, 192
Conflict of interest, 173–75
Confucius, 80, 130
Consequences of actions
    exterior, 39–42, 89–95
    interior, 54–55, 106–9
Contract model for the professional-client
    relationship, 178–79

Control over actions, 124
"Cooking" data, 160
Cooperation in evil, 111–13
  formal, 112
  immediate material, 112
  mediate material, 112
Copyrights, 163
Cost-benefit analysis, 224
Craft, of ethics, 35–36, 144, 241
Cynicism, 240

Dante Alighieri, 97
Deception, harm from, 144–46. *See also*
  Lying; Truth
Deontology, 150n.2, 238, 244nn.5–6
Descartes, Rene, 159
Developmental models of ethics, 21,
  28nn.7, 15, 253–54. *See also* Moral
  development
"Devil" problem, 128–29
Disclosure
  in contracts, 174
  of information, 178–79
Distributive fairness, 205–6
Distributive justice, 205
"Dolly" (the cloned sheep), 257
"Do no harm," 111, 211
Dow Corning Corp., 42–43, 60–61
Drugs, experimental testing in humans,
  227–28
Drummond, William (of Hawthornden),
  69
Duties, moral, 56, 111. *See also* Deontology

Egoism, 237
Einstein, Albert, 143
Emotions
  and ethics, 20, 236–37, 253–54
  component of psyche, 20
Employer-employee relationship, 191–92
Employment, social aspects, 191–93
Engineering
  distinguished from science, 157–58
  underlying principles, 161–62
Environmental concerns, 188–90, 192–93
Erikson, Erik, 254
Ethical complexity. *See* Complexity, ethical
Ethical serial, xiv
Ethics
  computer, 129–30, 148–49, 164–65
  as a craft, 35–36, 144, 241
  distinguished from morality, 9
  importance of, 4–5
  scope of, 9, 124
Ethics systems. *See also* Rationalist ethics;
  Virtue theory
  casuistry, 238–39
  choosing, 71, 241
  deontology, 238

developmental, 21, 28nn.7, 15, 253–54.
  See also Moral development
  egoism, 237–38
  four-component model, 257
  intuitionism, 238
  minimalism, 56–57
  monism, 239–40
  pluralism, 241–42
  postmodernism, 240–41
  psycho-social, 254
  relativism, 239–40
  rights-based, 238, 244n.7
  static, 253
  virtue based. *See* Virtue theory
  utilitarianism, 97n.2, 238
Event trees, 37–43
Expertise, technical
  and paternalism, 190–91
  and risk, 225–26
Exterior acts, moral analysis by, 36–42,
  88–95

Fairness, 23, 173
  in authorship, 175–76
  in conflict of interest, 173–74
  in contracting with clients, 178–79
  in employment, 191–92, 212–13
  and the environment, 188–90
  in expert advice, 178–79, 190–91
  and intellectual property, 187–88
  person-to-person, 173
  qualitative vs. quantitative, 174
  social, 187
  in supervising, 176–77
  in team projects, 175
Fair use, 164
Falsification of data, 160–61, 174
Fermat, Pierre de, 228n.1
Fishwick, Marshall W., 243
Food and Drug Administration, 227
Formal cooperation in evil, 112
Fortitude, 23, 78, 178, 257
Four-component model of ethics, 257
Fowler, James, 254
Frankena, William, 243n.2
Freedom. *See* Voluntary actions
Freud, Sigmund, 159

Gibran, Khalil, 43
Gilligan, Carol, 244n.8, 255
Glazer, Sarah, 150n.2
Goethe, Johann Wolfgang von, 58, 124, 213
Gossip, 147
Gould, Roger, 254
Grade inflation, 113–14
Growth, moral. *See* Moral development

Habits, 257, 260nn.21–22
  and choice, 23–24

development of, 55–56
and virtues, 22
Hammurabi, Code of, 111
Harmes, Joseph, 150n.3
Harvey, William, 161
Hellman, Lillian, 203
Helminiak, Daniel, 243n.1, 259n.2
Hierarchies
of values, 72–74
of virtues, 78
Hobbes, Thomas, 243n.3
Holmes, Oliver Wendell, 35
Hooker Chemicals, 7
Hus, John, 157

Imanishi-Kari, Thereza, xiii
Immediate material cooperation in evil, 112
Imprinting on character, 54–55, 106–9,
    114n.1
Indifference, 53
Informed consent, 57–58, 227–28
Intellectual property, 162–63, 187–88
Intention. *See also* Imprinting
of the action, 67n.1
in analyzing ethical complexity, 106–7
balancing against exterior consequences,
    110–11
definition, 52–53, 67n.1
importance of, 36–37, 53–55
and moral responsibility, 124–26
multiple, 52–55, 106–7
of the person, 67n.1
Interior goodness of actions, 106–7
Interior intentions, moral analysis by,
    36–37, 53–54, 114n.1. *See also* Intention
Internet
censorship of, 148–49
illegal copying of music from, 164–65
Intuition, 238
Intuitionism, 238
Involuntary actions, 21, 124–26
due to fraud, 124
due to invincible ignorance 124
due to passion, 125
impediment to freedom, 124–25

Jaki, Stanley, 259n.14
Jemez Mountains, 242–43
JTEX geological experiment, 242–43
Justice, 23, 78, 143
distributive, 205
and the environment, 188–90
qualitative vs. quantitative, 180n.2

Kant, Immanuel, 141, 150n.1, 244n.5,
    259n.12
Kiflik, Arthur, 165n.12
Knowledge, ownership of, 162. *See also*
    Intellectual property

Kohlberg, Lawrence, 254–56
Kubie, Lawrence, 160

Lasswell, Mark, 150n.3
Levinson, Daniel, 254
Lewis, C. S., 228n.2
Lichtenstein, Sarah, 229
Likelihood, 97n.2, 98n.3
and consequences, 88–89, 106–10
cumulative, 89–90
history of theory for, 228n.1
and risk, 223–25
Limits on moral responsibility, 125, 128–29
Line-drawing, 72–73, 125–26, 239
Locke, John, 244n.7
Loevinger, Jean, 254
Love, kinds of, 78
Love Canal, 26–27
Lowrance, William, 228n.3
Lying, 145, 150n.1

Martin, Mike W., 229n.9
Mathematical analogies in ethics, 88–89, 107
Mean, the virtues as a, 23
Mediate material cooperation in evil, 112
Melanism, 244n.9
Mental illness, 125
Mental reservation, 145
Mere, Chevalier de, 228n.1
Merit, allocation by, 207
Merton, Robert, K., 159
Methods, ethical. *See also* Ethics systems;
    Principles, ethical
differing, 237–42
selecting, 71–72
using, 35–36
Mill, John Stuart, 244n.4
Models
for the person, 19–21
properties of good models, 19
Monism, ethical, 239–40
Moral, distinguished from nonmoral, 11n.1,
    73–74, 97n.2
Moral development, 23–24, 35–36, 55,
    253–54. *See also* Developmental Models
    of ethics
Moral duties. *See* Duties, moral
Morality, distinguished from ethics, 9
Moral responsibility. *See* Responsibility,
    moral
Moral systems. *See* Ethics systems
Moral training, 22
MP3 technology, 164–65
Murray, William, 141
Mysticism, 256

Natural law, 150n.1
Need, allocation of resources by, 208–9
Negligence. *See* Comparative negligence

Net goodness, 89–90
Newkirk, Ingrid, 74, 80n.1
Newton, Issac, 165n.3
Nietzsche, Friedrich, 243n.3
Nihilism, 240
Nonmoral. *See* Moral, distinguished from
     nonmoral
N-rays, 190

Objective morality, 8–9
Obligations. *See* Duties, moral
Occidental Chemical Corp., 26–27
Occult compensation, 24–25
Ovid, 27

Paradigm case, in casuistry, 258–59
Pascal, Blaise, 228n.1, 253
Patents
   described, 188
   strategic, 188
Paternalism
   in business contracts, 190–91
   model for the professional-client
     relationship, 178
Peer review, 179–80
Penn, William, 114
Person, model for the, 19–21
Philo, 69
Philosophy
   defined, 7
   and fundamental ethical principles, 6–8,
     24, 36, 56, 111
Piaget, Jean, 254
Plagiarism, 160–61
Planck, Max, 159
Pluralism, ethical, 241–42
Poincaré, Henri, 187
Polyani, Michael, 259n.14
Polywater, 190
Postmodernism, 240–41, 244n.13
Prejudice in employment, 191–92
Principles, ethical. *See also* Methods, ethical
   consequences of selecting, 71
   differing, 237–39
   fundamental, 8–9, 24, 36, 56, 111
   justification for, 8–9
   need for, 7–9
Privacy, 147–48
Probability. *See* Likelihood
Professional-client relationships, 178–79
Proportionality in moral decision making,
     124–25, 174–76, 226
Protagoras, 244n.10
Prudence, 23, 57, 78, 178, 257
Psyche, 20
Psychology, 21, 125, 236–37, 254
Public Health Service, 57
Publishing, scientific, 158–59, 175–76,
     179–80
Pythagoreans, 80n.4

Random allocation of resources, 209–10
Rationalist ethics, 253–57, 260n.22
   advantages, 254–55
   developmental models, 253–54
   disadvantages, 255–56
   static models, 253
Relativism, 239–40, 244n.10
Religion
   defined, 7
   and fundamental ethical principles, 6–8,
     24, 36, 56, 111
Research, 57–58, 188. *See also* Authorship,
   order of in publication
Resources
   defined, 205–7
   negative, 206
Resources, allocation of
   by ability to pay, 209
   by merit, 207
   by need, 208–9
   random, 209–10
   by social worth, 207–8
Responsibility
   degrees of, 125–26
   moral, 9, 124–30
   of scientists and engineers, 5, 192
Rest, James, 257
Richmann, Georg Wilhelm, 221
Rights-based ethics. *See* Ethics systems,
   rights-based
Right-to-know laws, 164
Risk, 189
   acceptibility, 225–26
   defined, 222
   and informed consent, 227
Risk-benefit analysis, 224
Rotblat, Joseph, 258
Rowe, William D., 229n.8

Safety. *See* Risk
"Sainthood" problem, 128–29
Scandal, 112, 150n.4
Schinzinger, Roland, 229n.9
Schweitzer, Albert, 221
Science
   distinguished from engineering, 157–58
   underlying principles of, 158–59
Self-interest, 56, 237
Seneca, Lucius Annaeus, 259
Shamanism, 237
Shaw, George Bernard, 149
Slander, 145
Smith, Wolfgang, 259n.14
Social worth, allocation of resources by, 207–8
Sociopathy, 256
Socrates, 1, 244n.5
Software engineering, 129–30
Solomon, King, legend of, 114n.3
"Solomon" problem, 111
Sophists, 240

Space shuttle *Challenger*. *See Challenger* spaceship
Stages of moral growth. *See* Ethics systems, developmental
Static models of ethics, 253
Strategic patents, 188
Stumpf, Samuel Enoch, 244n.10
Superfund, 192–93
Surveillance of employees, 148

Tacitus, Cornelius, 165
Tarantino, Quentin, 244n.12
Tartaglia, 228n.1
Technology. *See* Engineering
Temperance, 23, 78, 178, 257
Ten Commandments, 244n.6
Teresa of Avila, 28n.15
Theological virtues, 28n.16
Therac-25, 114
Thrasymachus, 240
Three Mile Island, 189
Trade secrets, 163
"Trimming" data, 160
Truth, 23, 143–44. *See also* Deception; Lying; Privacy; Whistleblowing
harm from unnecessary spreading, 147
and intellectual property, 162–64
and judgment, 160–61
person-to-person, 143
in physical law, 157–58
in scientific practice, 160
social, 157
Turner, Clark, 130
Tuskegee syphilis experiment, 57–58

Unconscious mind, 21, 125
Union Carbide Corp., 96–97
Utilitarianism, 238
compared to method of present book, 97n.1, 114n.2

Values
of engineering, 161–62
hierarchies of, 78–79
moral, 78–79
nonmoral, 78–79
of science, 158–59
Virtue
connection to personal effort, 55
defined, 22–24
theological, 28n.16
theory, xiv–xv, 71, 97n.2, 239, 257
Voluntary actions, 9, 124–25

Weckert, John, 150n.5
Weighing. *See* Balancing
Whistleblowing, 146–47, 150nn.2–3
Work, 191–93

Zeno the Stoic, 19